Assorted Dimensional Reconfigurable Materials

Edited by Rajendra Sukhjadeorao Dongre and Dilip Rankrishna Peshwe

Published in London, United Kingdom

IntechOpen

Supporting open minds since 2005

Assorted Dimensional Reconfigurable Materials
http://dx.doi.org/10.5772/intechopen.77790
Edited by Rajendra Sukhjadeorao Dongre and Dilip Rankrishna Peshwe

Contributors
Rajesh Sreekumar, Marta Michalska-Domanska, Pazhanivel Thangavelu, Bhuvaneswari Kandasamy, Palanisamy Govindasamy, Bharathi Ganapathi, Lourduraj Stephen, Tahira Mahmood, Madeeha Aslam, Rajendra Sukhadeorao Dongre, Dilip R. Peshwe, Abdul Naeem

First published in London, United Kingdom, 2020 by IntechOpen
IntechOpen is the global imprint of INTECHOPEN LIMITED, registered in England and Wales, registration number: 11086078, 7th floor, 10 Lower Thames Street, London, EC3R 6AF, United Kingdom
Printed in Croatia

British Library Cataloguing-in-Publication Data
A catalogue record for this book is available from the British Library

Additional hard and PDF copies can be obtained from orders@intechopen.com

Assorted Dimensional Reconfigurable Materials
Edited by Rajendra Sukhjadeorao Dongre and Dilip Rankrishna Peshwe
p. cm.
Print ISBN 978-1-78985-513-5
Online ISBN 978-1-78985-514-2
eBook (PDF) ISBN 978-1-83968-193-6

For EU product safety concerns:
IN TECH d.o.o., Prolaz Marije Krucifikse Kozulić 3, 51000 Rijeka, Croatia, info@intechopen.com or visit our website at intechopen.com.

Meet the editors

Rajendra S. Dongre received his M.Sc. from the Department of Chemistry, R.T.M., Nagpur University in 1996 (Gold Medalist) and his PhD in 2010. His research work includes organic synthesis, chitosan bio-composite, assorted dimensional matrix, and remediation of water pollution de-fluoridation; nitrate, chromium, and phosphate lead (II). He has worked as a Scientist-B in the CSIR-LAB, National Environmental Engineering Research Institute (NEERI) Nagpur M.S., India. Overall, he has 25 years of experience in research and development and 18 years of post-graduate teaching experience, which has resulted in 70 international research paper publications. He has guided four research students to pursue their PhD. He received the 6th National Award (runner-up) for Technology Innovation in Petrochemicals and Downstream Plastics Processing Industry, for research in the field of polymer science and technology, handed by the Honorable Ananth Kumar, Petrochemical & Fertilizers Minister of Government of India in 2016. He received the 5th National Science & Technology Award for research contribution in the field of developing science in 2017, by EET-CRS, Noida, India.

Dr. Dilip R. Peshwe is a HAG Professor of Metallurgical and Materials Engineering and Dean (Faculty Welfare), Visvesvaraya National Institute of Technology (VNIT), Nagpur India. His areas of research interest include physical metallurgy, the welding and casting process, tribology, polymers, composites and wear of engineering materials. Eighteen students have completed their PhD under his guidance. He has published more than 150 research papers in international peer-reviewed journals and is the author of four books and six patents. He is the recipient of various prestigious awards such as the Jawaharlal Nehru Memorial Trust Award, Best S&T Innovation Award of Khadi & Industries Commission, and the IIM-SAIL Gold Medal of Indian Institute of Metals.

Contents

Preface

Since the beginning of history, many substances have been used as the embryonic form for progressive techniques and technologies through advanced materials in assorted fields such as polymer science, composites, electronics, membranes, nano-materials and bio-medicals. Despite advanced technology and scientific developments, many challenges are emerging and these are mainly regarding the reinforcement of material matrixes. Novel technologies explore the innate complexity involved in making assorted dimensional reconfigurable materials with superior biocompatible functions and interdisciplinary applications. Inspired by nature, material scientists are continuously engaged in designing and fabricating assorted multifunctional composites/matrixes for the advancement of science and technology. An intense outlook of the hierarchical structures or natural functions to bring about changes in materials for deriving precise applications is emerging in advanced multifunctional composites. This book focuses on the integration of potentially relevant bio-mimetic properties and/or functions such as self-cleaning with anti-reflection, self-cleaning with anti-corrosion, self-healing with self-cleaning, shape memory and wet-ability, super strength and toughness, and self-healing with super toughness. This overview outlines potential applications of some multifunctional composites/matrixes in the following industries: biomedical, oil-gas, aviation, marine, wearable electronics, fabrics, sensors, energy harvesting devices, and many more. As the frontiers of technology widen, assorted dimensional materials are created through physico-chemically reconfigured alterations, which deliver significant outputs and earn great attention in last few decades. Today, material for sustainable and green development in Science and Technology draws noteworthy global interest. Certain emergent materials are mainly reconfigured owing to paramount importance in industry and technology such as polymer-based sensors/markers, thermal conductors, metal alloys, piezo-energy harvesters, thermoelectric, and anticorrosive materials. This book contains six chapters written using vast research from diverse areas of material chemistry and the book describes the fundamental methodology, practices, and applications.

Introduction

Today's emerging nanotechnology approaches aid in making assorted dimensional reconfigurable materials for progressive development and advanced engineering functions in modern S&T. Reconfiguration augments new functionality in the resultant multiphase matrixes such as composite, blend, and hybrid. This introductory chapter is a preamble to interactive materials for the basic skeletal reconfiguration to derive specific applications. Rational designing and systematic reinforcements at various material scales are fundamental to manipulations native limiting characteristics and functionalities, which are unattainable through usual/non-reinforced materials. Skeletal reconfiguration in materials can be decisive to fabricate the particle, thing, and device at the atomic and molecular dimensions being achieved through assorted multidisciplinary fields such as biotech, physics, chemistry, plastic, polymer, and nano-engineering. Reconfigured/reinforcement reduces innate spatial dimension or captivity within the crystallographic phase of material by virtue of altering its physical, mechanical, thermal, optical, electrical, and electronic properties. In general, reinforced material matrixes contain three nanoporous skeletons, namely: 3-D/zero dimensional (e.g., particle, grain, shell, capsule, ring, and colloidal), 2-D/one dimension (e.g., quasi crystal, nanorod, filament, tubes and quantum wire), and 1-D/two dimensional (e.g., disc, platelet, ultrathin film, super lattice, and quantum well).

This chapter cited many smart matrixes yielded through rational designing of flexible materials skeletons for innovative findings in many domains including sensor, photo-detector, LED, laser, FET, physics, catalysis, biomedicine, environmental, aerospace, and construction. The introductory chapter overviews assorted 1D/2D/3D matrixes, which appeared to be best choice for soft and hard composites besides modernization of many fields including bio-electronic/optic, drug therapy, medical product, tissue engineering, smart battery, super-capacitor, electro-catalyst, bio-adsorbent, and manufacturing. While going through this chapter and the entire book, researchers will be aware of the superiority of reconfigured matrixes by virtue of designed features such as better load reassignments, tailored interfaces, increased strength, heat protection, and impactful solidity viable at low cost with end-use components. In a nutshell, reconfigured material matrixes are promising and notable due to novel advancements and a myriad of utility in futuristic S&T.

2-D Graphene Based Matrix

Energy is the crucial and universal input to drive life/routine cycles of human and nature. Conventional energy generating sources are nonrenewable as the native reserves are depleting extensively in addition to fossil-fuel burning that causes serious challenges such as global-warming and environmental pollution, so in this perspective energy generated through green resources is inevitable. Non-conservative renewable energy resources (such as solar and wind) are inexhaustible and abundant and this leads to a viable solution for such environment threats. In the light of the serious issue of energy generation through green sources, the research group of *Tahira Mahmood* presented a good summary of the newest progress in 2-D graphene reinforced nano-metal oxide based composite to be used as emerging feedstock for making photo-anodes in 3G advanced solar-cells such as dye-sensitized solar cells (DSSCs) and perovskite solar cells (PSCs). While discussing their investigation data, considerable attention was given to potential turn-key solutions to address burning challenges involved in making cheap and efficient photo-anodic substitutes for usual platinum electrodes. Many approaches are discussed in this respect, including the superior materials obtained through graphene matrix in designing novel counter electrodes in 3rd generation solar cells. Amid new materials with potential utility in promising 3G solar cells, graphene derivatives are worthy due to unique features such as conducting anodic potential and electron donor and acceptor characters. The authors have discussed the driving factors such as increasing charge transportations and reducing charge recombination viable for enhancing the performance of graphene nanocomposite as a photo-anode/electrode in DSSCs and PSCs solar cells.

3-D Metal - Non-Metal Matrix

Lourduraj Stephen's chapter has involved research to improve performance of 3-D metal-non-metal matrixes obtained through nano-technologically induced skeletal reinforcement. Recently varied metal/non-metal matrixes are fabricated with distinctive features such as super-porosity, non-biodegradability, flexible shape/size, and great surface area viable for superior applicability. In this regard, the author has controlled morphology of 3-D nanotitania to yield 7.7 nm dimension particles being smaller than commercially offered nanotitania of 25 nm ranges. This chapter gives an overview of titanium dioxide nano-crystalline material with particular porosity, pore-size distribution, and nature of pores being modified via in-situ synthetic conditions. The author has optimized certain preparative parameters in the making of 3-D nanotitania films apt to deliver superior performances in optical, structural, and electrical fields alongside photo-anode usage in solar cells due to native photoelectric/photovoltaic significance.

Marta Michalska-Domańska's chapter describes the surface-sensitive enhancement of Raman scattering by molecules being adsorbed onto nanostructures and how enhancement is sufficiently able to detect even a single molecule. Surface-enhanced Raman spectroscopy (SERS) strategically offers broad applicability in various fields, yet native progress is limited due to substrates used in SERS and their impeding characters such as less sensitivity, stability and repeatability with high reproducibility. Commercially available SERS substrates have an enhancement factor in the range of 10^5–10^7 besides possessing partial durability by virtue of little control over the substrate's morphology at the nano-scale. In this overview, the author has summarized certain anodic oxides derived nano-composite substrates to be used in SERS analytical technique. The preferred substrates for SERS analysis must have unique features such as homogeneous nano-structures, large surface areas, and self-cleaning to perform analysis with utmost accuracy and precision. The author has exposed many challenges involved in improvement of SERS substrates besides focusing on minimizing native drawbacks. Nowadays porous anodic oxides are obtained through cheap and scalable methodologies; this chapter cited some progressive SERS substrates and briefly compared these with conventional substrates.

Thankam Sreekumar Rajesh's chapter investigated Al_2O_3-40% TiO_2 powder blend for atmospheric plasma spray coating onto SS316 stainless steel to be used in manufacturing of jigs, guides, shafts, and fixtures. They utilize the teaching learning-based optimization (TLBO) algorithm to optimize input parameters and to find the best output parameters beside conducted confirmation tests, which gave the same values as predicted without significant error. Three levels L18 orthogonal array (OA) design of experiments (DoE) is applied to conduct this research study. The main input parameters measured included: nozzle distance, substrate speed, arc-current, carrier gas flow, and coating powder flow and the output parameters involved were: coating thickness, micro-hardness surface roughness, abrasion rate, and % porosity. Further mathematical models were made for each output parameter. Analytical hierarchy process was used to get weights for unit output parameters with generated objective and combined objective functions through the TLBO algorithm (global optimum values of input parameters and all the output parameters obtained simultaneously). The TLBO algorithm is easy and helpful to solve many multi-objective optimization problems.

2D-2D Metal - Non-Metal Matrix

The research group of *K. Bhuvaneswari* intercalated organic matrixes such as 2-D graphitic carbon nitrides ($g\text{-}C_3N_4$) into LDHs to get a blend of organic matrixes being used as a photo-catalyst for degradation of methylene-blue dye from aqueous solution. g-C_3N_4 has unique semiconducting features such as large and tunable band gaps beside apt salts intercalation best suitable for photo-catalytic actions. Layered double hydroxide (LDH) is a 2-D inorganic ternary nano-composite putting trivalent cations within brucite-like layers with net positive charge onto surfaces. LDH is lamellar metal hydroxide materials consisting of carbonates and water intercalated in interlayer galleries. 2D organic gC_3N_4 is intercalated into 2D inorganic LDH imparts frame heterojunctions, which improved photo-catalytic activity of nitrogen-rich g-C_3N_4 intercalated LDH towards direct degradation of dye through photo excitation, oxidation, and reduction by visible light absorption. The authors claimed to have advanced photocatalytic activity due to more charge separation rate and huge photo-induced electrons collectively leading to effective degradation of methylene blue dye.

The perspective involved in each chapter is scholar, meticulous, and unambiguous for which the editors are genuinely thankful to all contributing authors for making their efforts instructive as well as practicable. This book will direct multidisciplinary researchers to recognize aspects of assorted dimensional reconfigurable materials for progressive advancement of S&T.

Dr. Rajendra Sukhjadeorao Dongre
Department of Chemistry,
RTM Nagpur University,
India

Dilip Rankrishna Peshwe
Professor,
Department of Metallurgy,
VNIT,
Maharashtra, India

Acknowledgements

I take this opportunity to heartily thank all those who lent hands in any way for finishing this book project. First and foremost, I am thankful to Co-editor, Prof. Dilip R. Peshwe, Head, Professor of Metallurgical-Materials Engineering, VNIT, for a whole-hearted co-editorship job.

It gives me immense pleasure to recognize the keen and genuine contributions of all authors that made this book a reality. I am grateful to Ms. Jasna Bozic, Author Service Manager, IntechOpen publisher for sustaining incessant contact between authors and editors.

I am indebted and obliged to each of the below mentioned people who have had a huge impact in my academic, research, and professional career for their backing, encouragement, and support, no doubt being precious, graceful, and highly blissful.

Prof. R. B. Kharat Ex-Director, Institute of Science, Nagpur-22, MS. India.

Prof. K.SN. Munshi Ex- Head Department of Chemistry, Nagpur University, Nagpur, India.

Prof. M.G. Paranjape Ex-Head Department of Chemistry, Nagpur University, Nagpur, India.

Dr. D. S. Ramteke Ex-Head EIRA-Division, NEERI, Nagpur-440020, MS., India.

I pay special gratitude to family friend Shamim Sheikh and well-wishers Aslam Bhaij, Nagpur for their encouragement, inspiring words, and appreciations.

Last but not the least; I express deep gratitude to my adorable parents, beloved wife Ulka, angel daughter Isha and my siblings for their invaluable care, love, and priceless moral support. Above all, I thank the Almighty for the unlimited showers of blessing.

This book is dedicated to late Sir Stephen William Hawking, a great physicist, cosmologist and Director of the Research Centre for Theoretical Cosmology, Cambridge University, UK.

Section 1

Introduction

Chapter 1

Introductory Chapter: Assorted Dimensional Reconfigurable Materials

Rajendra Sukhjadeorao Dongre and Dilip R. Peshwe

1. New trends in futuristic 2D material matrixes: to graphene and beyond it

The reconfigured matrixes begin a crucial research domain, as it is a need and necessity of new materials to crack challenges in domestic, food-packing industry, environment, computer, engineering and technological for twenty-first century and beyond [1]. Functional material matrixes play a major role in tackling such confronts, solution to the harmful environmental pollution, for light-weight aerocraft components, sustainable constructions, energy-generation/storage, food-packaging, and space journey [1, 2]. Finding of materials endure non-stop R&D as growth and usage of novel multifunctional materials is quite vast. Identifying the worth and prospective of functionality for the progress of futuristic materials, global researchers ever explores many materials milieu. Without doubt, lots of fields apprehend merit of functional reinforcements in materials to derive varied utility including nanotechnology, shape memory, ferroelectric, electronic, thermal, conductor, insulator, opto-electric, magnetic, phase-change and biology [1–3]. Science and technological progressions were persistent from past few millenniums and amplify with great tempo in twenty-first century. In this advanced nanotechnology era discovery of more practical and sustainable novel materials are significantly augmented.

Profuse 2D materials can capably alter into splendid matrixes in practical mode with innate development in nano-scale and atomic-level applications [4]. Graphene's 2D skeleton has inspired great interests in reinforcement of many 2D/3D templates famed as recent "alchemy" attempts to modify all possible periodic table elements into creative and applied matrixes. Such reconfigured 2D/3D matrixes endowed diverse devices, tools and gadgets leads to finer quality and superior optical encoder outputs for assorted industrial usages. Yet, core stability and large size restraints applicability of graphene, which can be overcome by some changes like functionalization and substrate-based reinforcement in 2D frameworks. 2D/3D materials seek special innovations to correct limiting features through synthetic reconfigurations due for device-production approach like hetero-structure advancements being practical for novel applications and opportunities [5]. Introductory chapter is an overview of recent avenues for 2D/3D matrix owing assorted alterations, reconfigurations and designing to get diversified applications in advancement of S&T in twenty-first century.

IntechOpen

2. Modern matrixes in development of S&T

New matrixes owing manipulated modern functionalities yield through reconfigurations offer progressive applications in S&T. Many comprehend materials endure in matrix reinforcement through precise and rational designing can endow novel characteristics as missing in usual materials. Modern science and nano-technology assisted diverse reconfigurations in material skeleton so as to produce assorted nano-materials, decisive-particles, species and devices both at atomic and molecular levels [5]. Logical reconfigurations in material cut down its spatial dimension in local crystallographic phases via augmented features including mechanical, physical, chemical, thermal, optical, electrical, electronic and rheological. Many reconfigured matrixes own mentioned nano-porous frameworks as zero dimensional/3D (particle, grain; shell; capsule; ring; colloidal), one dimensional/2D (quasi crystal, nano-rod; filament; tube; quantum wire) and two dimensional/1D (disc; platelet; ultrathin film; super lattice; quantum well/dot). Nowadays, many 1D/2D/3D matrixes like graphene, germanene, silicene, carbide, nitride, MXene, spintronic, etc. are employed in creation of advance devices and tools as reconfigured through respirocyte, nano-dendrimer micelle, drug conjugate, carbon nanotube, quantum-dot/well [1–6].

3. Reconfigured magnetic 2D materials beyond graphene

Graphene is mere a tip of iceberg and finding of optional 2D materials like metal oxides, metal hydroxides and chalcogenides and metal-organic framework is the opening to spot rest of whole iceberg. Magnetic materials hold significance in diverse fields such as; data storage, electronic, and bio-medical. Many 2D materials like, h-BN, metal dichalcogenide, metal hydroxide and carbon nitride are reconfigured via attenuated features like strain, void, defect/vacancy, tangible magnetism, doping, adatom, dangling and bond induce low-dimensional magnetism [5, 6]. Certain layered material like $CrXTe_3$, CrI_3, trisulfide and 2D metal oxide such as MoO_3, $Ni(OH)_2$ beside perovskites viz.; $CaTiO_3$ ($^{XII}A^{2+}\,^{VI}B^{4+}X^{2-}{}_3$) are reinforced through unusual strain-and-layer govern anisotropic magnetic ordering [4–6]. A few framed matrixes hold extra degree of freedom called valley state are best reinforced with precise tasks like spintronics, and photoelectronic potential to be employed for fast processing and huge data storage next-generation devices [6]. Diverse reconfigured 2D matrixes show quasi magnetism induced between few-atom-thick layers, e.g., MXene, metal-organic framework, metal carbide, nitride and carbo-nitride. Lots of synthetic reinforcement are feasible in 3D lattice which consumes innate crystalline imperfections like interfacial-defect, interstitial ion, substitution impurity, imperfection, dislocation in order to tackle the exigent dimensional and structural changes.

4. Metal-organic frameworks (MOFs)

Metal organic framework is a matrix reinforced with organic linker amid metal node own unique features like, huge surface area, great porosity, tuneable pore-size and lithe functionality leads to various modern and multi-functional utility in S&T [1–6]. MOFs advanced usages includes gas-storage, oil-water separations, hetero-catalysts, sensors, proton-conductivity and biomedicines. Past few decades have reconfigured many matrixes to establish strong perceptive in buildup of progressive MOFs for desired utilities. Advance MOFs are formulated as periodic, convenient,

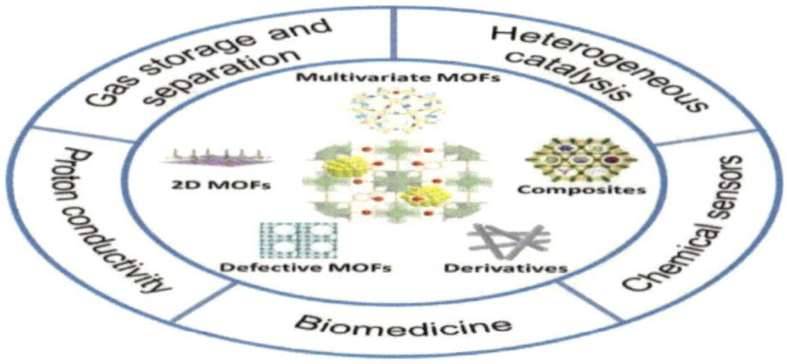

Figure 1.
Possible utility of metal-organic frameworks (MOFs) in S&T.

nano-scaled matrixes owing large specific surface area from single/grouped metal which leads to diverse applications including organic strut, linker, ligand, sensor, marker, gas/energy-storage, CO_2 confiscation, electro-catalysis, and drug delivery besides filtration, oil-spillage/radioactive sludge clearance. Reconfigured MOFs have prospective tasks in tackling critical issues in the future era. MOFs have creditable control on movement of one moiety in concern to another as best for exploitation in separation and catalysis beside own innate ability of holding precise molecules being practicable for chemisorptions, high and low-pressure adsorption testing besides storage. Such capacities in MOFs are derived honestly through reconfigured internal skeletons, so careful and precise investigations get invoked for progressive matrixes offering characterized functions. MOFs owe unique features like surface strengths and molecular interactions need to reconfigure for native functionality like hydrophobicity, hydrophilicity and superior catalytic activity. Possible utility of MOFs in S&T as shown in **Figure 1** includes testing many gases like NOx, H_2S and SO_2 and volatile organic compound with control repeatability.

5. Formulated liquid metal matrix

Binghamton University, USA have formulated such liquid metal lattices embraced mutually via silicone covering which crushing/heating get back to its native form [4]. Many liquid metal matrixes are reconfigured to discover myriad applications including soft optoelectronics, liquid metal robots, foldable antennas and aerospaces, etc. Caltech Institute, California, have developed amorphous liquid-metal alloy called Vitreloy (trade name) in 2003, for making industrial things, golf-club, watch and cell-phone covers [6]. First liquid metal lattice was reinforced in rubber shell as fields alloy from bismuth, indium, and tin metal leads to superior usages like; portable/grid energy-store, rechargeable battery electrodes nuclear plants/reactor coolant, 3D printing, vacuum casting and electronic circuitry as shown in **Figure 2**. They owe pied characteristics like high tensile strength, deformability, corrosion resistive, electronic conductance, superior electrochemistry and anti-wear capacity thus provides conformal coat/guard against humidity, dirt, chemicals and temperature. Some liquid metal matrixes are amorphous at NTP and induce heat during its processing thus workable as substitute to thermoplastics [7].

Some liquid metal matrixes are safe, sturdy and suck huge energy on crushing, besides regain usual shape after heating/cooling, thus reuse without further processing as shown in **Figure 3**. Many liquid metal matrix tenders great prospect to NASA satellite missions, space-crafts and rockets as designer can group "spider web" into tiny package being easy to open out as transmitter in rotator orbits [6, 7] (occupy less area onboard vessel, expand on lands at target). Scientists fabricate

Figure 2.
Room-temperature liquid metal alloy matrix for flexible battery/energy store.

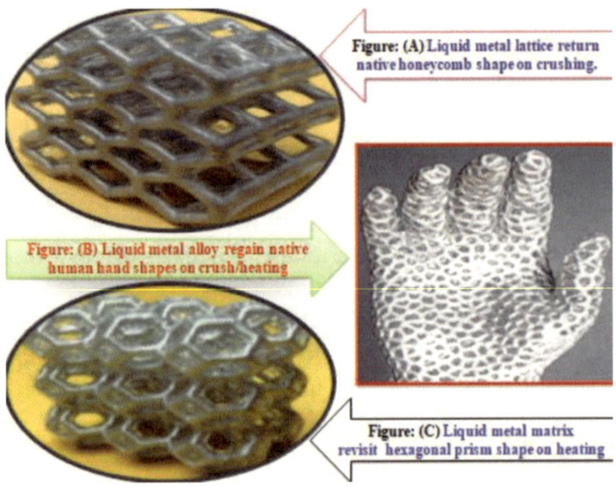

Figure 3.
Some liquid metal matrixes/lattices retrieve native shape on crushing/heating.

an interplanetary ship using designed liquid metal matrixes inbuilt cushion owing spacecraft with crashing chance on planetary landing, since liquid metal absorbs energy and gets deformed but regains innate shape on heating later can be reused [8].

6. Notable reinforcements in 2D graphene

Graphene own long strands of carbon, thin than hairs are stiff and strong which is often used in making stronger synthetic carbon fibers [9]. Experimental studies and computer simulations have confirmed significance of matrixes reconfigured through lightweight, vigor material called 2D graphene carbon-fibers. Although, carbon-fiber is expensive, its reinforced matrix imparts superior safety in fabricated products like light weight-cars, equipment, gadgets and mainstay of aeroplanes. Trace graphene amount 0.08% w/w reinforced carbon fiber offers 225% larger strength and 184% greater stiffness over PAN-derived fibers. Thus, needs to exploit viability of graphene in making advanced and reconfigured carbon fibers from cheaper precursors, in order to reduce production cost even less. 2D graphene reinforcement produces robust fiber matrixes owing high strength and low manufacturing cost, since usual method uses costly polyacrylonitrile/PAN which enhances 50% cost due to energy inputs [9]. Graphene's planar topology aids constitutional alignment all through the resultant carbon fiber, helps reconfigured matrix to squeeze around its edges. Graphene is the strongest known 2D material, but many 3D structures are reconfigured owing unique features such as highly porous, 10 times strengthen than steel, and quite lighter spongy composites/matrixes are achieved through advanced nanotechnology techniques. 2D graphene owes robust skeleton/matrix owing many incredible characteristics, except its innate thinness is incompatible in ensuing 3D matrixes. But, this graphene is prone to diverse artificial alterations in geometric configurations due to reinforcements of assorted materials (like synthetic/natural polymers, metal/non-metals and inorganic) that yields varied 3D matrixes.

Rice University, USA have developed laser-induced graphenes/LIG which shoed labeling ability onto many edible material surfaces including toast, coconuts and potatoes. Multiple laser induce graphene-foam seize cross-linked 2D carbon flake aids marking materials thus leads unique function in numerous fields like supercapacitors, fuel-cells electro-catalyst, RF-recognition antenna and bio-sensors/markers. LIG can turn into paper, cardboard, cloth, coal, and foods beside acts as tag/sensor for *E. coli* bacteria detection. Laser-induce graphene (LIG) foams can reinforced with polystyrene, plastic, rubber, cement, wax materials which tendered smart and robust packaging as significant in many field applications like wearable/flexible electronics, heat-therapy, water cleansing, anti-ice/de-ice shell, antimicrobial surface design and resistive random-access memory devices. LIG in laser burns thin polyimide sheets and yields intercalated graphene flakes which appeared alternative to woods. Many 3D matrixes are sculpted through reinforcement of 2D graphene foam aids blending with many functional composites/bio-composites and emerged vital module to create dynamic objects in electronic, ultra-hydrophobic medical equipment and textiles.

LIG layered 20 μm-thickness electrodes kill bacteria and prevent microbial fouling through antimicrobial action demonstrate several applications like water treatments, hospitals and seawater pipes exemplify surfaces as liable to bio-fouling [10]. LIG reinforced organic additives yield composites expand the range of green and sustainable applications like biomedical films, nanogenerators, puncture detectors, de-icing/anti-ice coats, flexible heating pad and pre-command heat-up garments. Graphene reinforcement in powder sugar and nickel yields foam which is further use to develop novel objects/feedstock for 3D printing/imaging. 2D graphene reinforced carbon nanotube yield 3000 times load supportive 3D rebar framework which is ultra-strong, conductive and protect innate shape best for designing aircraft, battery de-icing nanoopto-electronic, and tissue/bone-implant materials. Laser sintered technique reinforces nitrogen/sulfur in graphene to yield fingertip-block rebar foams carrying 3D matrixes show use in energy storage/damp and sound absorptions [10, 11].

6.1 Reinforced quantum matrixes

Firm quantum matrixes are reconfigured to avail innovative opportunities were unknown earlier [12]. Several metal alloys show surprising performance for traversing potential in 'spin liquid states' as investigative mapping of such quantum materials criticality has established a traveler traverse the final frontier. Here, traveler is metallic-alloy owing ordered constituents following parallel paths which gets altered with applied pressure, temperature and magnetic field, thus knowing concern electron's behavior. Certain *reinforced quantum material matrixes* own self-assembled, tuneable interfaces as observed in plotted journey in their order to disorder pattern. Anti-ferromagnetic interfaces cross solitary border in adjacent zone and marked as paramagnetic materials due to harmony amid trillions of electron amending mutual position. Some reinforced matrixes follow such quantum criticality and it is easy to analyze relative phase changes. Liquid metal alloy acts like of spin liquid, though metallic stress yields quantum matrix. Reconfigured metal alloys are used to study weird electronic excitation that aids designing of high-temperature superconducting quantum materials. Assorted self-assembled, tuneable interfaces are reinforced in quantum materials which have revolutionized modern device developments and thrust innovative highly adoptable electronic devices afar present imagination. In fact quantum materials are more complex than conventional semiconductors; thus lay great task in designing of clean interfaces through advanced designing. Research has revealed incredibly improved and novel functional characteristics arise amongst interfaces of reinforced quantum materials.

Advance techniques yield diverse liquid metal-alloy/matrix that leads to offer designed native electrical features being at par/superior to familiar semiconductors. Certain reinforced metal matrixes have shown incessant shift from metal to semiconductor on varying thermodynamic conditions such as density, admix proportion/ratio and temperature. Some synthetic matrixes like liquid selenium and Cs-Au liquid-alloy are develop with switch metallic-semiconducting performance due to altered native chemical interactions/configuration. Liquid gallium is reconfigured via tri-block organic-copolymer as void elastomeric matrix offering superb electric conductance besides solid to liquid phase changes on small melting being best feedstock to yield ultra-stretched shape memory fibers. Reinforcement induces solidified core transformation in liquid-gallium imparting better stiffness and great deform shape with fiber modulus change (from 4 MPa to 1253 MPa). Elastic energy hoard is seen in such hollow shell of liquid gallium during deformation which relaxes elastomeric fiber and regain usual shape on small heating, thus endowed shape memory features. In fact gallium is used to make shape memory fibers due to many valuable features like; good metal framework, improved electric/heat conductance, ideal fixity and adjustable effectual modulus. Low melting alloys are reconfigured as fusible matrixes permitting metal to reshape into liquid/semi-solid state at low temperatures with subsequent re-solidification. Malleable and elastomeric conductive fibers are reconfigured for opted rigidity and variable shape memory modulus which offered outstanding utilities in flexible electronics, soft robotics, and wearable devices. Many reinforced fibers are crafted in complex geometry via twisting and angular parting which provides deformation with preserved electrical continuity in resultant matrixes. The qualitative properties of material controls reinforcement of matrix and direct designing of various parts, equipment and products for industrial/practical usages. Throughout matrix reinforcements the constituting materials get adjoined and casted to design templates without collapsing further. Nano-technologically brings lithe reinforcements even at molecular level by stirring strong electrical interconnecting targets, so lay basis for assorted matrixes to be used in flexible electronics, light-weight aerospace components and robotics.

Some reinforced matrix seizes magnetic moment allied in unlike directions but parallel to interfaces (unconcern to close-set electrons) lack magnetism due to shielding/screening. Theoretical modeling is use to reinforce complex competitive interactions at quantum levels in such reinforced matrixes. Neutron spectroscopy also illustrated such interactions in metal matrix viable for impulsive pattern wherein electronic and magnetic characters get swapped periodically. Periodic array can attend interfaces amongst swapping layers akin to interfaces seen in reconfigured hetero-structure. Auto/self assembled interfaces attain through quantum engineering impart clean in-situ interface-width directed through applied magnetic field and temperature. Many reconfigured matrixes shown remarkable features like retrieve energy absorption, adjustable rigidity and elasticity and gifted various advance applicability in designing web-mesh antenna, aerospace, soft-robotic,

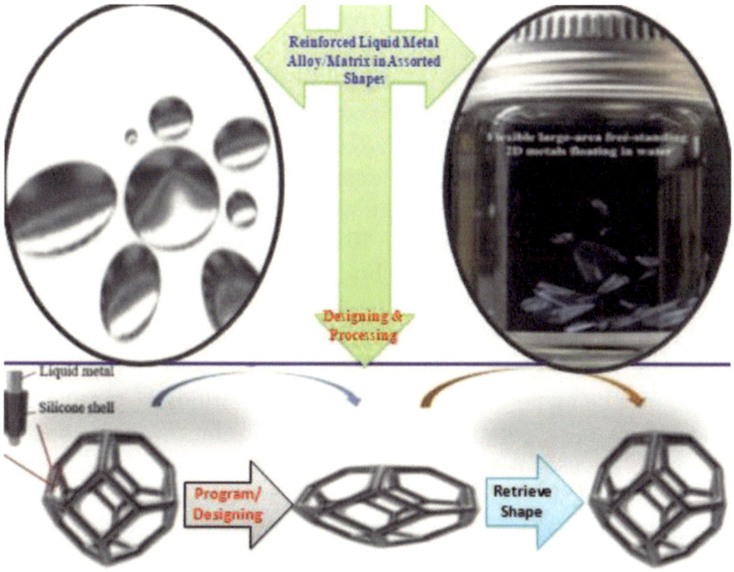

Figure 4.
Multifunctional liquid metal matrixes yield through hybrid design/built-up.

meta-material, 3D printing, flexible electronic and coolant for nuclear reactor [13]. Many multifunctional liquid metal matrixes/alloys we obtained via hybrid design/built-up method as shown in **Figure 4**.

7. Specially reconfigured 2D metals

2D matrix owes single-layer pattern of crystalline materials owing distinct physical and chemical features leading to assorted applications like photovoltaic, semiconductor, electrode, water-oil separation/purification, marker, bio-sensor, etc. [14]. Amid 2D matrixes self-supporting metallic matrix is very hard to obtain due to involvement of characteristics 3D structural bonding. Top-down/bottom-up synthetic approach produces self-supported single-layer 2D metal matrixes owing surface controlled properties and stability <2 nm cross-section, but giant films size and range are limited. Wet-chemical paths yield bulk 2D metal matrix, while few atomic layer reinforcements are obtained via mechanical exfoliation (metal's tiny plane size is lesser than few micrometer). Indeed it is very tricky task to create bulky 2D metallic frameworks as large as and as chemically complex as 3D matrix. Facile reinforcements are obtained through advance nanotechnology are best employed in mechanical devices. Vapor deposition technique has re-configured coating of thin metallic layers at the apex of hydro-gel substrates yielding swallow/deform exfoliated films. In-plane dimension/chemical composition method has reconfigured much precious freestanding 2D nanomembrane without physical margin viz.; Ti-metallic films, more entropy alloy FeCoNiCrNb and metal-glass ZrCuAlNi, non-layered ceramics, semiconductors, polymers, composites [15]. These reinforced 2D matrixes owe 3D chemically complexity and pave a path to unknown remarkable world of low-dimensional matrixes lead to novel usages including soft robotic, flexible electronic, filtration, bio-composite and bio-engineering. Certain low dimensional self-supportive metallic/non-metallic membranes are reinforced for restricted plane size <10 μm, but polymer exfoliated folding yields apt geometry/morphology as chemically complex as 3D matrixes owing surface induced physic-chemical characteristics as shown in **Table 1**. Such freestanding material matrixes gained terrific interests due to distinct nano-skeleton and shrink dimensionality, special physico-chemical features and indent modern applications. Reconfigured 2D metal system hold inherently bonding akin to 3D scale, thus emerged as thermodynamically stable atypical single layer (cross-size <2 nm) matrixes. Tiny 2D metal matrixes are tricky for practical applications, while steady high surface-volume reinforcement augment innate thickness from 0.8 nm to 50 nm and lateral dimension from 20 nm to 8000 nm with aspect ratio from 10 to 1000, thus defeats thermodynamic stability barrier. Despite geometry/morphology restraints reconfigured 2D metal matrix showed unusual features like eminent electrical/heat conductance, great flexibility and proactive surface which is best for flexible electronics and clean energy production. Reinforced 2D Au/Ag-Bi metal matrixes shows profound surface plasmon resonance being best for LSPR-sensors applications.

Bottom-up mechanical compression, nanolithography, top-down solution-base chemical methods are used in reinforcement of 2D metal alloy and glass in bulk amounts. Assorted synthetic reinforcements are used to direct geometry of 2D metal matrixes as nano-sheets, membranes and films. Masking agent also controls film deposition and results desired shape 2D surfaces like Ag-C matrix, ultra-thin nano-TiO_2 sheet, silicon nano-sheet, high entropy alloy, ZrCuAlNi, and SiC ceramic.

PolySlide Company yields reinforced composite-tubes and optional metallic feedstock to be used in making of pneumatic and hydraulic cylinders. PolySlide tubes are blend of filament-glass fiber with resins due for unusual aspects like dynamic component, non-conductive, good dimensional stability, non-corrosive and impingement resistance beside sustainability at extreme conditions viz.; high/low temperature, oil/grease, grit, salt and acid/base chemicals. Such reinforced cylinder owes size range from 0.25-in. inner diameter till industrial grade 24-in. diameter. PolySlide tubing found to prevent galling which let sealing slide over applied surface contour and reduces knit friction. Some reinforcements are derived through frameworks like wound glass-fiber, resin and polytetrafluoroethylene in many matrixes viz.; composite tubing, bearing and pneumatic cylinder. Such matrixes are preferred over metals due to unique features like huge load-bearing capacity, little friction, non-corrosive, resistant, self-lubricant and no-greasing. Reinforced matrixes reinstate metal usages for making special things like building equipment, mining/farming wheels, compact track loader, excavator, back-hoes implement, applicator, cart and spreaders [15].

Reconfigured methods	Metal used	Thickness (nm)	Lateral size (nm)	Aspect ratio
Confine growth method using organic ligands	Palladium	0.8	100–200	100–200
	Silver	2	500	250
	Copper	2	20	~10
2D milieu-confined growth	Gold	2.4	200–500	80–200
Hydrothermal	Rhodium	0.4	500–600	1000–1500
Nano-particle assembly	Platinum	10	50–100	5–10
Mechanical compression	Aluminum	2	7300	~3500
Exfoliation	Antimony	4	1000	250
PSBEE based 2D process	Titanium dope high alloy/glass, FeCoNiCrNb	10–50	10^4 to 10^7	10^2 to 10^6

Table 1.
Relative analysis of thickness and in-plane dimension of 2D metal matrixes yield via different methods [13, 14].

8. Reincarnated quantum states in 3D materials

Several 3D superconducting quantum materials are reincarnated as electronic "energy stacks" being active at 2D topological surfaces [13]. Every recreated energy stack own special states viable for 2D quantum Hall effect viable for mesoscopic transportation in quasi-1D matrix through maintained parity and time-reversal symmetries (controlled by energy gain and loss). Quantum movement in reincarnated matrix is analyzed by non-Hermitian system only in unbroken phases (not for broken phases) in the energy band using exceptional points. As broken phase allowed spontaneous symmetry states wherein cross-stitch matrix is separated into two identical single lattices equivalent to degenerate eigen states. In quantum phases of matter, intrinsic interfaces exist amongst electrons as quite complex than classical atomic and molecular interactions [16].

Michael Faraday study of light and matter interaction onto thin gold leaf showed faint ruby color fluid production was the first ever reported quantum states, too supported its validity [13–16]. Today quantum physics and computing have certified such quantum things a bit more irony, as formerly noted as unbelievable. Superconductivity is observed in many such reincarnated quantum matrixes such as quark-gluon, simple plasma, degenerate matter, Bose-Einstein condensate and quantum-spin liquid. However, quantum spin liquids are surprising due to never aligned but continuously oscillating electron spins even at the lowest absolute zero (as normal matter's spin get frozen). This quantum spin liquid is valuable and practical reincarnated electronic matter at quantum scale due to superior electronic applications and quantum mechanical impacts. Modern S&T have reconfigured many 1/2/3D materials like ballistic $LaAlO_3/SrTiO_3$ conductors, quantum hall phased graphene-$SrTiO_3$ matrix and magnon stretch graphene being gifted with unique features like planned quantum state, regimented magnetic stimulations and unique electrical conductance [17]. Advanced nanotechnology generates magnon spread viable for quantum Hall effect or ferromagnetism across the bulk path all over graphene reinforced magnetic phases due to survival of 2D excited electron waves at low temperature. Scientists has reinforced prolong spin-wave guides in-out of quantum hall edge channel via electro-magnetically regimented quantum states across insulated anti-ferromagnetic phases of 2D graphene.

In twenty-first century, many properties of quantum nature of matter were either ignored or unknown earlier are ever explored due to endowed applications in many quantum fields including computing, sensing, teleportation, and communication [1, 13–17]. Reincarnated quantum states offer electric potential with fine-tuned state marking transition to topological order along with paternal quantum entanglement as viable for quantum Hall effect through such reconfigured 2D/3D matrixes. Scientists have reincarnated remarkable quantum Hall effect in topological superconductors to be used in futuristic fault-tolerant quantum computers. Reincarnated stable state matrixes own quantum entangled storage and process information which aids creation of quantum computing. Physicists had discovered 2D/3D phased topological materials for quantum hall effect and then reincarnated "energy stacks" of 2D electronic states in 3D superconductors for designing of quantum computing.

9. Reconfigured matrixes for spintronic

Spintronics is a gist of spin transport electronic which engaged in study of electron spin-fields allied magnetic moments as cogitated in semiconductor based electronic transistors and metal derived solid-state devices, since from 1990 (multiferroics akin in insulator) [18]. Spintronic vitally differs from usual electronics, since spin and charge are used as extra degree of freedom for data storage and transfer. Spintronic effect is mainly comprehended in thin magnetic semiconductors and certain magnetic alloys (2:1:1 copper, manganese and tin). Certain opticomagnetic matrixes are reinforced for spintronic including Heusler-alloy like Co_2Mn-Si and more band-gap doped semiconductors like ZnO, TiO_2 to be used in quantum and neuromorphic computing [5]. Spintronic matrixes reconfigured solid-state devices perform electron transportation through spin-fields and charges at quite speedy and inexpensive way due to less energy usages in spin transformations than current production. Manganese-gallium and arsenic non-oxide alloy has designed for unique interfacial resistance and tunnel barrier induce ferromagnetism utilized in quantum and neuromorphic computing. Dzyaloshinskii-Moriya interactive spin orbit torques results robust interfacial magnetic skyrmion through many reinforced polycrystalline/amorphous matrixes like Pt/CoFe/MgO, Pt/Co/AlOx CoFeB and Ta/CoFe/TaOx with chirality govern vertical anisotropic superior ferromagnetism. Novel antiferromagnetic matrix controls skyrmion Hall dynamics and aids certain designing criteria like low damping, great thermal stability, larger voltage ruled magnetic anisotropy, writability at low cost, readability and data storage/transfer being vital in making racetrack memory based skyrmionic devices [19]. Such reinforced matrixes offer lots of incentives in getting high density memory based advanced devices through designed big spin Hall angle and electric field exchange MRAM phenomenon.

10. Graphene for spintronic devices

Graphene offers huge platform for designing spintronic based devices due to native augmented spin-dependent electron transportation with control spin-current effect [20]. Spintronic device own spin-field channels obtained via atomic interactions of randomize electrons thus cart spin signals over transistor-scale. Reconfigured matrixes like Si-Ga-As and graphene generates such field-channels to cart signals with balance energy split amongst up and down spins [19]. But graphene offers remarkable features including big spin-time, strong spin-orbit precession, controlling spin-field swap, suitable spin relaxation path (sinking spin signal integrity) and compel equilibrium being vital in designing wise spintronic circuits/devices. Graphene reinforced cobalt matrix yield electrodes offer good contact and enhance nonlocal spin valve and superior spin transportation without any variation in spin-signals in micro-scale devices [20]. Due to cobalt scattering such graphene derived electrode own unique qualities like collateral interactions, reduce diffusion path and spin-relax alterations which aids carting of spin-signals upto 5 mm distance amongst all tested materials. Graphene-boron nitride reinforced matrix bids superb mechanical, optical, thermal, electric and magnetic properties like huge specific surface area, elevated Young's modulus, duel contact over an air-gap and reduce substrate interactions via self-supported layers being useful for lots of advancements in many fields including photo-electricity, catalysis and transistors. Rise of graphene has modernized technology and structural designs in spintronic/spin-current base devices/circuits including micro-electro-mechanical system (MEMS), energy-harvesting systems, sensors, actuators, ink-jet printers and flexible electronic as scalable silicon alternative. Such graphene reinforced matrixes have also replaced carbon-based nano-electronics and spin-based devices besides ferromagnetic and atomic switches. 2D graphene gave many advanced and novel dimensions to the material matrixes including nano-ribbon and nano-sheets wherein layers shrinks width < 10 nm owing band gap suitable in fabrication of transistor circuits. Assorted technological reinforcements endowed large surface active properties and non-zero band gaps thus makes graphene more beneficial than CNTs in MEMS/new electronic-device in last decade. Graphene derived all-spin logic-single micro/nano-electronics create whole spectrum of spintronic devices (low energy usage) beyond existing technologies. Reinforced graphene matrixes grow many novel applied technologies striking new market and applicability like spintronic and magneto-resistant hard disks are game changers in electronics and optoelectronics by.

10.1 Reconfigurable 2D/3D conjugated polymers

Conjugated polymers enter a new aspect in 2D material matrixes due to unique applicability in flexible electronics and soft robotics. Certain structurally reconfigured 2D conjugated polymers pave innovative growth in material science by virtue of superior electronic parameters [19, 20]. These conjugated polymeric reinforcement yields altered lattice symmetries and structural properties leads to progressive utilities of functional materials beyond graphene. Expertise domains of sciences like organic chemistry, condensed matter physics and nano-materials have teamed up so as to get special designed reconfigured 2D/3D material matrixes offering grand features as guessed hypothetically until now. Polyaniline (PANI) polymer reinforced 3D matrixes are developed for energy storage and induced rate electricity transport in especial gadgets like mobile, electric vehicles and power-grid [21]. PANI reinforced conductive electrodes and supercapacitors stores charge by exchanging electrons with ion through dopant material thus results in large energy storage through "pseudocapacitance" effect. Nano-carbontube is reinforced into PANI matrix to owe capable high energy densities designed for super-capacitor to be used in power storages. Futuristic global energy exploits such reconfigured matrix leading technologies for intermittent storage and high-power delivery devices for effective and broad potential applications.

10.2 Graphene's cousins abet dream to reality

Carbide, nitride, silicene, germanene, stanene, black phosphorus and Zintl phase materials are few known 2D reconfigured matrixes called graphenes cousins offer gifted properties seek for spintronic and quantum computing [19–23].

11. Reinforced polymer/dendrimer matrixes: current and future developments

Reinforced polymer based matrix are futuristic smart materials discovered for the prospective advancement in S&T. Nano-metal doped polymeric matrixes are best to be used as fillers due to designed parameters such as shear stress, shape, size, control-rate, concentration and time dependent non-Newtonian mechanics, shear-thinning and visco-plastic flow controls. Metal reinforced nano-materials own significant tuneable features including rheology as suitable for deriving advanced utilities in photo-voltaic, catalysis, optics, drug career, smart adsorbents and energy-storage devices. Bottom-up, self-directed assembly of polymer block is powerful mean for robust reconfiguration and well manipulation of desired matrixes. Nano-material based polymeric matrixes display amazing characters due to inbuilt atom-by-atom reinforcements being absent in usual bulk counterparts. Assorted dendrimer designing are achieved through click chemistry and nanotechnology, currently discovered promising reinforced polymeric matrixes for progressive S&T. Novel nano-periodic systems acquired for dendrimeric reinforcements through material templates in the form of nano-devices, nano-materials, and nano-medicines. Many commercial products are architecture with designed characteristics via advances dendrimeric reinforcements. Such dendrimer reinforcements transmit ideal nano-structure obtain during repetitive branched monomer iterative protection and de-protection. Rationally reconfigured dendritic matrixes are employed in drug delivery, electro-catalysis and light emitting materials [1, 2]. This novel dendrimer bid base for organic-inorganic hybrid pass on diversified utilities. Hyper-branch dendrimer forms 3D globular, viscous and highly soluble matrixes owing superior features like low-viscosity, high solubility and multi-functionality over linear-branched counterparts. Such reinforced dendritics are beneficial due to innate parameters like elastic, light weight, high resistance, processable, robust, chemical inert and thermal stable. Hyperbranched dendrimeric matrixes hold linear segments with spherical end being adaptable in fabrication of nano-composites like polyamidoamine dope-nanogold, nanocarbon-sols and heterogeneous nanostructures [3]. Click chemistry carried reinforcement aids stimulus detachment in porous dendrimeric links and yields various matrixes including layer-by-layer films, nano-particles, nano-sheets, nano-wires, and nano-tubes and optoelectronic nano-devices [2–4]. Reinforced dendritic matrixes showed peculiar characters like linear/cross-link/chain branching, open functionality; classified topology, co-hybridization and terminal grafting. Flexible dendritic matrixes offered innovative chemical and biology applications in many industries. Dendrimeric polymer owns quite handy size/shape and compartmental zones made-up through hyper-branch and tree-reconfigure templates capable of surface functionalization, mono-dispersions with immense stability make it smart carrier for drugs. Dendrimer derived nano-device have latent value in cancer chemotherapy without stimulated immune side-effects since targeted drug therapy through designed tectodendrimer matrixes (dendrimer unit exhibit assorted role in targeting, disease diagnosis, drug career and imaging) [2]. Boron EGF-reinforced PAMAM matrix own drug targeted cells preferably in intra-tumoral injection. CED-doxorubicin-2,2 bis (hydroxymethyl) propanoic acid reinforced dendrimers dipect in vitro less toxicity in colon carcinoma cells [1–3]. Cationic dendrimers owe more cytotoxicity, cell membrane instability and cell-lysis over anionic dendrimeric-PAMAM matrixes [1–4].

12. Concluding remarks and futuristic developments

2D graphene reinforced many pioneer matrixes have established practical developments in modernization of S&T beside thrusts novel domain of many 2D materials including layered double hydroxides, LAPONITE-clay, boron nitride, black phosphorous, graphitic carbon, carbide, nitride, heavy metal-halide/oxide, perovskite-oxide and polymer [1–4]. Reconfigured matrixes gift unique features like; flexible skeleton, mechanically strong, facile electronic /thermal transportation, zero band-gap, semimetal nature and highly mobile electron driven conductance. Numerous matrixes feed basic need of S&T besides promoting researchers to widen a plethora of 2D/3D materials. Fascinatingly, broad domains of assorted 2D/3D materials are developing everyday and more than 150 remarkable matrixes are reconfigured till date [1, 23]. These matrixes endowed surprising technologies, and innovative findings in many domains including sensor, photo-detector, LED, laser, FET, physics, catalysis, biomedicine, environmental, aerospace and construction. Patented superior outputs at nano-scales and low energy consumption are the key encouragements in electronics hunt for device architectures through such reconfigured 2D/3D materials [1].

Since graphene discovery, scientists have found a series of 2D/3D layered matrixes through reinforced physical and chemical features and thus tackled functional restrictions of 2D materials. Graphene reinforcements aid to replace half of its carbon with other extrinsic atoms and able to derive diverse stacking in 2D/3D matrix/hetero-structure/hybrid owing new/designed properties [2, 23]. Many reconfigured hetero-structures own special features as missing in graphene/ other materials and best heir to most popular silicon in electronic devices like solar panels. Scientists are devising optional matrixes to cater skilful demands in energy storage to fix silicon with competent materials. Assorted bulk 2D/3D matrixes of the counter materials are being explored for innate beneficial optoelectronic features due to practical applications at economic scale beside substitutes for outdated technology [1–4].

This introductory chapter is an overview of assorted 2D/3D material matrixes which appeared as alternatives for soft-hard composites and modernize many fields including bio-electronic, optics, drug-therapy, medical product, tissue engineering, battery, super-capacitor, electro-catalyst, adsorbent and aerospace component manufacturing. Intrinsically reinforced matrixes yield through diverse materials found superior than their counterparts by virtue of designed features like better load reassignments, tailored interfaces, strong, heat protection, and impactful solidity viable at low cost with end-use components. Reconfigured material matrixes are promising due to notably attenuated performance utilities and persuade novel advancements in futuristic S&T.

Author details

Rajendra Sukhjadeorao Dongre[1*] and Dilip R. Peshwe[2]

1 Department of Chemistry, RTM Nagpur University, Campus, Nagpur, Maharashtra, India

2 Department of Metallurgy, VNIT, Nagpur, Maharashtra, India

*Address all correspondence to: rsdongre@hotmail.com

References

[1] Dongre RS. Chitosan-derived synthetic ion exchangers: Characteristics & applications. In: New Trends in Ion Exchange Studies. Vol. 1. Croatia: IntechOpen; 2018. pp. 21-42. DOI: 10.5772/intechopen.78964

[2] Dongre RS. Marine Polysaccharides in Medicine. Croatia: IntechOpen; 2017. pp. 181-206. DOI: 10.5772/65786 ISBN: 978-953-51-2860-1

[3] Dongre RS. Rationally fabricated nanomaterials for desalination and water purification. In: Novel Nanomaterials. Vol. 1. Croatia: IntechOpen; 2018. pp. 348-366. DOI: 10.5772/intechopen.70149. ISBN: 978-1-78923-089-5

[4] Dongre RS. Nanomaterials via reconfiguration of skeletal matrix. In: Nanostructures. Croatia: IntechOpen; 2020. pp. 181-206 ISBN: 978-1-78923-994-2

[5] Giananglo B, Holst B. Surface Science Techniques. Springer; 2013. pp. 3-4. Chapter 1

[6] Goh GD, Yap YL, Agarwala S, Yeong WY. Recent progress in additive manufacturing of fiber reinforced polymer composite. Advanced Materials Technologies. 2019;**4**(01):1800271

[7] Fernandez JG, Dritsas S. Biomaterial age: The transition toward a more sustainable society will be determined by advances in controlling biological processes. Matter of Opinion. 2020;**2**(6). DOI: 10.1016/j.matt.2020.04.009

[8] Khan K, Guo Z. Recent developments in emerging two-dimensional materials & their applications. Journal of Materials Chemistry C. 2020;**8**. DOI: 10.1039/C9TC04187G

[9] Sethulakshmi N, Tiwary CS. Magnetism in two-dimensional materials beyond graphene. Materials Today. 2019;**27**. DOI: 10.1016/j.mattod.2019.03.015

[10] Qin Z, Buehler MJ. The mechanics and design of lightweight three-dimensional graphene assembly. Science Advances. 2017;**3**(1):e1601536

[11] Bondavalli P. Graphene, microelectronic/nanoelectronic & more than moore. Graphene & Related Nanomaterials. 2018. DOI: 10.1016/b978-0-323-48101-4.00006-0.

[12] Park S, Baugh N, Dickey MD. Ultra stretchable elastic shape memory fibers with electrical conductivity. Advancement of Science. 2019;**6**(21):1901579

[13] Ryu J-W, Myoung N, Park HC. Reconfiguration of quantum states in symmetric quasi-one-dimensional lattices. Scientific Reports. 2017;**7**(1):8746

[14] Wang T, He Q, Yang Y. The controlled large-area synthesis of two dimensional metals. Materials Today. 2020;**36**. DOI: 10.1016/j.mattod.2020.02.003

[15] Guo X, Zhang L, Guihua Y. Room-temperature liquid metal & alloy systems for energy storage applications. Energy & Environmental Science. 2019;**12**

[16] Datta S, Das B. Electronic analog of the electrooptic modulator. Applied Physics Letters. 1990;**56**(7). DOI: 10.1063/1.102730

[17] Briggeman M, Levy J. Pascal conductance series in ballistic one-dimensional $LaAlO_3/SrTiO_3$ channels. Science. 2020;**367**(6479). DOI: 10.1126/science.aat6467

[18] Kang W, Wang Z, Zhao HW. Advanced low power spintronic

memories beyond STT-MRAM. In: Proceedings of on Great Lake Symposium on VLSI-2017; ACM. 2017. pp. 299-304

[19] Wei DS, van der Sar T, Halperin BI, Yacoby A, et al. Electrical generation and detection of spin waves in quantum Hall ferromagnet. Science. 2018;**362**(6411)

[20] Young F, Shepard CKL, Kim P. Spin and valley quantum Hall ferromagnetism in graphene. Nature Physics. 2012;**8**. DOI: 10.1038/nphys2307

[21] Eftekhari A, Li L, Yang Y. Polyaniline supercapacitors. Journal of Power Sources. 2017;**347**. DOI: 10.1016/j.jpowsour.2017.02.054.

[22] Pizzini S, Gaudin JG, Thiaville A. Chirality-induced asymmetric magnetic nucleation in Pt/Co/AlOx ultrathin microstructures. Physical Review Letters. 2014;**113**(4):047203

[23] Bandi S, Peshwe DR, Srivastav AK. Graphene from discharged dry cell battery electrode. Journal of Hazardous Materials. 2019;**366**(15). DOI: 10.1016/j.jhazmat.2018.12.005

Section 2

2-D Graphene Based Matrix

Chapter 2

Graphene/Metal Oxide Nanocomposite Usage as Photoanode in Dye-Sensitized and Perovskite Solar Cells

Tahira Mahmood, Madeeha Aslam and Abdul Naeem

Abstract

Global energy shortage will be one of the most critical challenges in the next 50 years. Currently, over 80% of energy consumed is produced from fossil fuels, which is directly linked to global warming and environmental pollution issues. Environment-friendly renewable energy is rapidly gaining importance for the existence of human civilization. A leading source of renewable energy is the solar energy, which is inexhaustible and abundantly available. Solar cells that convert solar energy directly into electricity are drawing considerable attention as a potential turnkey solution to address these challenges. Several approaches have been made in this respect, including the development of better materials and the designs of new solar cell configuration and architecture. Among the innovative materials with potential application in emerging 3G solar cells, graphene and its derivatives such as GO, rGO and G/nanocomposite have been widely explored as transparent conducting electrodes, electron donor or acceptor materials and counter electrodes (CE). In this chapter, the use of graphene nanocomposites has been explored as an electrode material in DSSCs and PSCs. Recently, graphene/metal oxide nanocomposites have been widely used in DSSCs and PSCs and played a significant role in increasing charge transport, reducing charge recombination and thus enhancing the performance of solar cell.

Keywords: graphene, nanocomposite, metal oxide, photoanode, energy, DSSCs, PSCs

1. Introduction

1.1 Solar energy as an alternative of renewable energy

Energy and environmental problems such as pollution and global warming are the most outstanding challenges that humanity will face in the next 50 years. Lately, global energy demand primarily for conventional energy resources such as fossil fuel, coal and natural gas has been increasing with population growth and industrial development. Up till now, over 80% of energy consumption is met with from fossil fuels, which cause global warming and environmental pollution issues. Moreover, they are non-renewable and will ultimately be exhausted in the future.

IntechOpen

Due to the increasing interest in renewable energy sources, imminent research is being focused towards harvesting energy from natural resources. Among the various choices, exploitation of sunlight for energy production and environmental remediation is the utmost crucial research areas of the twenty-first century.

The ultimate renewable source of energy is the sun. Sunlight is the most abundant energy source available to mankind. The sun emits a large amount of energy (approximately 32 × 1024 joules/year), which is much more than the world's whole demand. It is abundantly available, and only its 0.01% use would meet the worldwide energy necessities [1]. In this context, photovoltaic (PV) technology is considered as one of the ideal candidates, as they convert solar energy directly into electricity. Hence, effectively converting solar energy directly into electricity has been a promising solution to the energy issues [2]. From the perspective of environmental protection and energy conservation, it is required to produce electricity from solar energy by means of photovoltaic devices or solar cells.

1.2 Solar cells

Solar cells are probably the foremost contributor to fulfill the future energy requirements. Several approaches have been made for the fabrication of solar cells. Generally, solar cells are classified as first-, second- and third-generation solar cells. The first-generation (1G) solar cells are also called conventional, traditional or wafer-based cells and include silicon solar cells (polysilicon and monocrystalline) [3]. The power conversion efficiencies (PCE) of crystalline silicon solar cells have reached up to 26.6%. The SSCs with more than 90% share are presently leading the commercial photovoltaic markets. However, they are characterized by a severe preparation conditions and high cost [4, 5]. The second-generation (2G) solar cells, thin-film solar cells which include direct band gap semiconductors, were then explored. These included cadmium telluride (CdTe), gallium arsenide (GaAs), copper zinc tin sulphide (CZTS) and copper indium gallium selenide (CIGS) thin-film solar cells [4]. Though thin-film solar cells have led to reduced cost as compared to silicon solar cell, they require high-temperature and vacuum vapor deposition processes. Furthermore, they mostly contain toxic and rare elements, thus limiting their widespread applications [6, 7]. To overcome these challenges, third-generation (3G) solution-processed solar cells have been developed, which include organic solar cells (OSCs), quantum dot-sensitized solar cells (QDSCs), dye-sensitized solar cells (DSSCs) and perovskite solar cells (PSCs) [8]. Several PV technologies are under continuous development to meet the world demand for energy; in this chapter, however, we will focus our discussion in the emerging PV devices, i.e. dye-sensitized solar cells and perovskite solar cells.

The solar cell performance can be measured by its power conversion efficiency (PCE), cost and stability. Solar cell efficiency is the "measure of its output power per unit intensity of the incident light". The basic measurement to control solar cell efficiency is current (I) and potential (V) measured over an external resistance and at incident light intensity. The I-V characteristics can be produced using a Keithley SourceMeter SMU instrument. The I-V measurements were carried out to get significant knowledge about parameters of the solar cell for cell's performance. The typical illumination condition used for measurement of solar cell is 100 mW. cm^{-2} (intensity of incident light), which is recognized as one sun or air mass (AM) 1.5. It is the sunlight that reaches the earth's surface through air mass at 42.8° angle. In laboratories, a solar simulator attains this standard illumination condition [9]. The photovoltaic parameters thus calculated are open-circuit voltage (VOC), short circuit current (I_{SC}), fill factor (FF) and power conversion efficiency. In 2016, the power conversion efficiency values certified by the National Renewable Energy Laboratory (NREL), the best research dye-sensitized solar cell, organic photovoltaic cell (OPV) and the perovskite solar cell, are 11.9, 11.5 and 22.1%, respectively [10].

1.3 Graphene oxide

Exploring innovative materials with tailored nanostructures and desired properties for energy application is a recent research area. To this end, graphene has distinguishing advantages over conventional nanomaterials, and substantial efforts have been made to utilize its valuable features for efficient energy devices. Carbon nanomaterials such as graphene and CNTs due to their abundance, low cost, good electrical conductivity and high chemical stability have been applied in 3G solar cells [11, 12].

Graphene is a single thick sheet of sp^2-hybridized carbon atoms organized in a hexagonal lattice [13]. Graphene, first discovered in 2004, has appeared as a rising star in material engineering. Geim and Novoselov winner of Nobel Prize in 2010 opened infinite new possibilities for graphene, and recently, around the world many efforts have been made to present graphene-related materials to many industries [14]. Graphene has extraordinary properties such as high electrical conductivity (108 S/m), good thermal conductivity (5000 W/mK), high surface area (2630 $m^2\ g^{-1}$) and electron mobility (250,000 cm^2/VS), excellent mechanical strength, room temperature quantum hall effect, tunable band gap and good biocompatibility [15, 16]. Graphene is extremely optically transparent material having absorption (<2.3%), transmittance (over 97.7%) and insignificant reflectance (<0.1%) [11].

Due to these interesting properties, graphene, GO and rGO have attracted increased popularity for the use in optoelectronic (LEDs, photodetectors, touch screens, etc.), energy conversion (photocatalytic water splitting, photoelectrochemical (PEC) water splitting and photovoltaic cells), energy storage devices (batteries and capacitors) and environmental applications (gas sensors, photocatalytic degradation of pollutants and heavy metal removal) [17–19].

Geim and coworkers first isolated a 2D single layer of graphene from graphite by peel-off method in 2004 at Manchester University, named as Scotch tape method [20]. This led to an explosion of interest, and several studies have been carried out on the structure and properties of GO. Therefore, a series of approaches have been used to obtain a high-quality and large surface area graphene oxide. These approaches are classified into two main types: the bottom-up and the top-down.

1.3.1 The bottom-up

The bottom-up approach is simple and is exfoliation of a graphene layer from a graphite. However, it requires high temperature and produces graphene with comparatively more defects than the top-down approach. Chemical vapor deposition (CVD), epitaxial growth on single-crystal SiC and carbonization are the representative bottom-up methods. CVD is a distinctive method used to synthesize large-area graphene sheets on metal foil substrates [20–22]. However, due to complexity, high temperature and cost of metal substrates, these methods are not widely used [22].

1.3.2 Top-down

The top-down method involves the formation of graphene oxide using carbon-based materials. The top-down method has advantages like high yield, ease of operation and solution-based processability [12, 23]. This approach involves the chemical exfoliation of graphite, thermal exfoliation, electrochemical exfoliation and chemical reduction strategy. The typical method used for synthesis of GO was developed by Hummer and coworkers, which involves oxidation of graphite by sulfuric acid and potassium permanganate [18, 20].

Graphene oxide (GO) is the important derivative of graphene, which can be synthesized directly from graphite oxide. GO comprises of a 2D network of sp^2- and sp^3-bonded C atoms. The exceptional atomic and electronic structure of GO opens opportunities for new functionalities. GO is a highly oxidized graphene sheet containing many oxygen-comprising functional groups like carboxylic acid, hydroxyl, epoxide and carbonyl groups. Due to the presence of these oxygen-containing functional groups, graphene oxide is easily dispersible in water and other organic solvents [16, 23]. These functional groups are outstanding reactive sites for various functionalization reactions, nucleation and growth of nanoparticles [24]. Reduced graphene oxide (rGO) is an intermediate phase of GO and graphene, possessing various oxygen-containing functional groups and lattice surface defects which cause the electrocatalytic site in metal nanoparticles. Therefore, rGO exhibited better performance than the fully reduced defect-free graphene [12].

GO acts as semiconductor having band gap of 1.7 eV at room temperature. GO is an ambipolar material because it can act both as n- and p-type conductors. Therefore, they can be used both as ETL and HTL, depending upon the oxygen-containing functional groups in their structure. Graphene oxide, due to its unique structure and properties, has shown many advantages over rare metals, fluorine-doped tin oxide (FTO) and indium tin oxide (ITO) in optoelectronics such as solar cells and light-emitting diodes as shown in **Figure 1** [12, 15, 25].

Graphene oxide tends to agglomerate because of van der Waals interaction between the graphene layers, preventing its application in several fields, though this drawback can be avoided by hybridizing graphene with CNTs, metal oxides and polymers [11]. To further explore the potential application of graphene-based materials, graphene nanocomposites are attracting more and more interest (**Figure 2**).

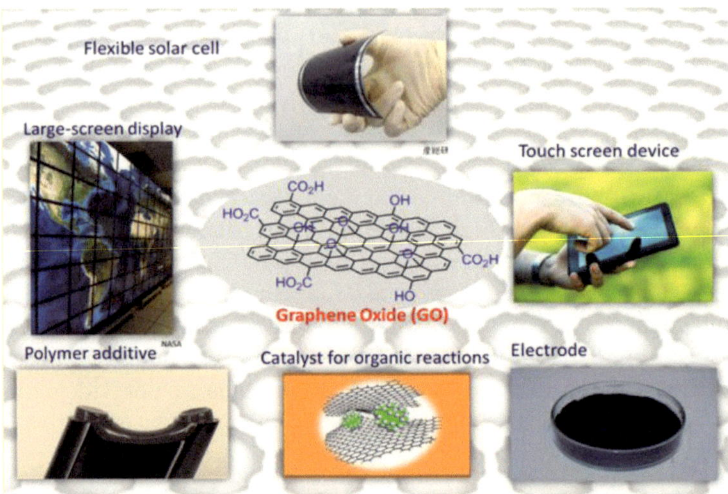

Figure 1.
Applications of graphene oxide in electronics.

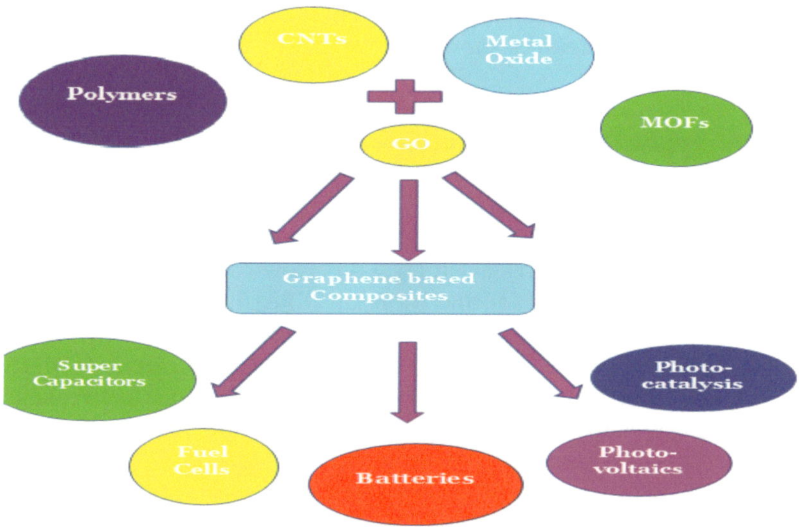

Figure 2.
Applications of graphene-based composite.

1.3.3 Graphene nanocomposites

Graphene nanocomposites are emerging as a class of exciting materials that hold promise for many applications. Generally, graphene-based composites are formed by incorporating polymer, ceramics or metal nanoparticles into graphene. The incorporation of such materials into graphene is very desirable for tuning the morphology of surface, electronic structure, and fundamental properties of graphene. The enhanced properties of the GO-based composites are because of excellent physical and chemical properties of GO and incorporated nanoparticles [26].

The functional groups of GO such as the epoxide, hydroxyl, carbonyl and carboxyl group offer the attaching points for nanoparticles. The larger surface areas and the conductive graphene structure facilitate the charge transfer and redox reaction and enforce the mechanical strengths of composites. Although graphene sheets naturally stack into multilayers and hence their high surface area and essential physical and chemical properties are lost, however, when nanoparticles are incorporated into it, graphene sheets can assist as support materials to anchor them and improve the properties of GO [24].

GO-based nanocomposites are the research hotspots nowadays due to their large-scale production and synergistically enhanced effect. Considerable efforts have recently been reported for decorating graphene with metal oxide nanoparticles. To date, several kinds of metal oxides such as TiO_2, ZnO, SnO_2, MnO_2, Co_3O_4, Fe_3O_3, Fe_2O_3, NiO and Cu_2O have been incorporated into GO [16, 27]. Nanoparticle growth on graphene sheets is a significant approach to produce nanocomposites or nanohybrids, as controlled nucleation and growth offer optimal chemical interactions and bonding between graphene sheets and nanoparticles, leading to very strong electrical and mechanical coupling within the nanocomposite. GO-metal oxide nanocomposites have been found as promising materials for lithium batteries, sensors, solar cells, fuel cells, photocatalysis and organic synthesis [16, 17, 24].

Recently, graphene and graphene-based nanocomposite have been extensively explored in emerging 3G solar cells particularly in DSSCs and PSCs. Graphene and its derivatives have been widely used as transparent conducting electrodes, electron donor or acceptor materials and counter electrodes, playing a substantial role in increasing charge transport, reducing charge recombination and thus enhancing the performance of solar cell.

2. Graphene application in solar cells

2.1 Graphene application in DSSCs

A dye-sensitized solar cell (DSSC or DYSC) is a low-cost solar cell, invented by Swiss scientists Michael Grätzel and Dr. Brian O'Regan in 1991, and is often called as Gräetzel cell (G Cell) [3]. DSSCs have attracted significant attention in search of substitute for silicon and thin-film solar cells due to their environmental friendliness, low fabrication costs and simple preparation process and remarkable conversion of solar energy into electricity [28]. DSSC is a photoelectrochemical system consisting of (a) transparent conducting working electrode (photoanode), (b) a dye sensitizer, (c) a counter electrode (cathode) and (d) an electrolyte [28, 29]. Its principle of operation is likened to natural photosynthesis process in plants. A monolayer of dye sensitizer absorbs incident light (like chlorophyll) giving rise to positive and negative carriers in the cell. On illumination, the electrons of the dye are excited from the highest occupied molecular orbital (HOMO) to the lowest unoccupied molecular orbital (LUMO). Then they are transported to the semiconductor's (TiO_2) conduction band (CB) and diffuse through the semiconductor, being collected at the transparent conducting oxide (TCO). An electrolyte solution which typically comprises of an iodide/triiodide (I^-/I_3^-) redox couple provides electrons to regenerate the oxidized dye. Eventually, the electrons transfer to the counter electrode (Pt) through the external circuit and reduces triiodide I_3^- ions back to the iodide I^- to complete the circle of conversion of photons into electricity (**Figure 3**) [12, 30, 31].

The photoanode is a key constituent, as it strongly influences the photovoltage (V), the fill factor (FF) and the incident photon-to-current conversion efficiency (IPCE) [12]. It serves as the main energy conversion center, converting photons into electrical energy, thus playing a critical role in DSSC [25]. A good photoelectrode is

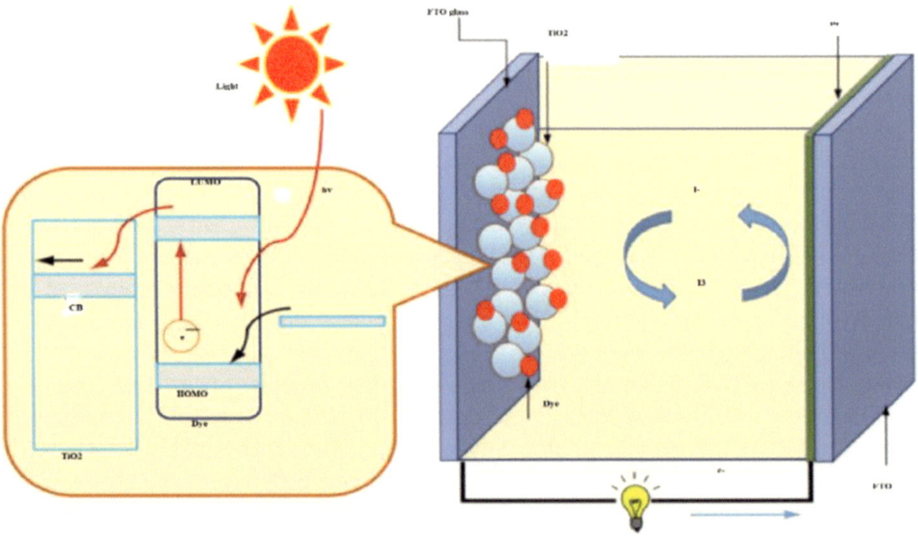

Figure 3.
Scheme diagram of dye-sensitized solar cells [12].

assumed to provide good usage of light, good electron injection and good electron collection. Various semiconducting (wide band gap) materials such as TiO_2, ZnO, SnO_2, CdS and Nb_2O_5 have been successfully used as photoanode in DSSCs [28, 32]. Among all, TiO_2 is most abundantly used, as it is nontoxic, cheap, chemically inert and easy to synthesize [33]. Moreover, TiO_2 have a wide band gap 3.2 eV; predominant chemical, electrical and optical properties; low recombination rate for the hole-electron pair; and great absorption property [12, 28]. The performance of DSSC is significantly influenced by properties like particle size, porosity and thickness of TiO_2 film.

Platinum (Pt) is the most widely used counter electrode (CE) in DSSC. Though Pt is very active to catalyze the reduction of triiodide, there are issues of stability in corrosive electrolytes and its high cost. Several efforts have been made to substitute it with less-expensive materials such as carbon materials and conducting polymers [34]. Carbon-based materials, such as graphite, carbon nanotubes, carbon black, nanocarbon and more recently graphene have been discovered severely as an alternative electrocatalyst for DSSC [17, 31].

Due to the exceptional mechanical, electrical and optical properties, graphene or highly reduced graphene are more appropriate as transparent electrodes, part of electrolyte, a light-harvesting dye and a counter electrode [17]. Additionally, work functions of graphene and rGO are calculated to be 4.4–4.5 eV, which are close to fluorine-doped tin oxide (FTO 4.4 eV) [25]. Thus, the tunable band gap of GO and its photon absorption ability in the visible and IR region permit to be a potential sensitizing material. The first application of graphene in DSSC as transparent conducting substrate was reported by Wang et al. [35] in 2008. They fabricated GO films by thermal reduction in Ar and/or H2 flux from exfoliated graphite oxide. These films showed a transparency of more than 70% and a higher conductivity of 550 S/cm [35]. Subsequently, other roles of GO have been widely explored in DSSC.

2.1.1 G/metal nanocomposite as photoanode in DSSCs

Among the extensively used semiconductors, TiO_2 possess the longest diffusion length and lowest charge effective mass, but electron-hole recombination remains a significant issue [30]. Hence in order to reduce electron recombination and improve the stability of DSSC, graphene and graphene-based materials as photoanode have received great interest since the past few years.

The idea of using graphene into the photoanode of DSSC arises from the pioneering work of carbon nanotubes mixed with oxide nanostructures by Jang et al. [36]. This led to open an interesting field for researchers, quickly extended to graphene and its derivatives. The fascinating properties of graphene such as high transparency and Young's modulus make it a good candidate for transparent conducting substrates. The semimetallic properties and ultra-fast electron mobility allow graphene to serve as a charge carrier and as a transporter inside the semiconductor oxide layer in photoanode.

Kazmi et al. [31] used graphene-TiO_2 nanocomposites as photoanode for DSSCs. They synthesized Gr-TiO_2 with different graphene concentrations. Graphene nanoparticles were synthesized by modified Hummer's method, whereas TiO_2 nanoparticles were purchased from Sigma Aldrich. The band gap was found to be reduced from 3.16 eV to 2.2 eV with the increase in GO concentration. Furthermore, conductivity increases with graphene concentration which supports well with the reduction in band gap. PCE of DSSC was enhanced with the addition of graphene, and optimum results were found for 3.0 wt% of graphene concentration.

Nouri et al. [30] have reported nanocomposite photoanode titania photoanodes with reduced graphene oxide rGO for DSSCs. They carried out comparative study on two ex situ and one in situ doping of TiO_2 with rGO. The structural, optical and electrical characterization of the TiO_2/RGO nanocomposites and the PV performance of the DSSC were studied by several techniques. UV-vis spectroscopy showed that the existence of rGO resulted in narrowing of band gap and visible light absorption, particularly in solvothermal in situ TiO_2/RGO, representing chemical bonding between graphene sheets and TiO_2 nanoparticles. This chemical bonding has been verified by XPS and Raman spectroscopy. In situ doping of TiO_2 with RGO had the greatest beneficial effects on the performance of DSSC devices, yielding the highest Voc, Jsc, η and IPCE values. This was primarily ascribed to the role of RGO into the TiO_2 films to facilitate electron transport and decrease electron-hole recombination.

Kanta et al. [10] carried out comparative study of the promoting effects of graphene in TiO_2 photoanodes. They investigated the effect of the types and concentration of rGO on structural and photovoltaic properties of TiO_2-based electrodes. GO was synthesized by modified Hummer's method. The synthesis of rGO was carried out by using two different methods: chemical reduction with vitamin C and thermal reduction. The thermal reduction method was conducted in situ during the fabrication and heat treatment processes of the DSSCs. It was observed that DSSCs containing GO/TiO_2 photoanode (rGO by in situ thermal reduction) showed greater photovoltaic performance than rGO/TiO_2. It was also found that the PCE of the DSSCs changed with the concentration of graphene in a nonlinear manner. The 0.01 wt% GO/TiO_2 showed highest PCE (3.69%) values whereas for rGO based DSSCs 0.03 wt% rGO/TiO_2 showed maximum PCE (2.90%).

Xu et al. [22] synthesized TiO_2-RGO nanocomposites via an ultrasonication-assisted reduction method using GO in the TiO_2 precursor solution. The reduction of GO and the formation of TiO_2 crystals were carried out simultaneously. TiO_2-RGO was characterized by SEM-EDX, FTIR, XRD, Raman, XPS, UV-vis and electrochemical impedance spectroscopy. By the introduction of RGO, absorption of light of octahedral TiO_2 was markedly improved. In the meantime, the PCE of DSSC containing TiO_2-RGO photoanode was also found to be highly improved.

Jamil et al. [28] fabricated DSSCs using G-Nb_2O_5 photoanode. Nb_2O_5 due to higher conduction band edge and open-circuit voltage is attracting attention. In order to reduce charge recombination and improve charge collection efficiency, an electrically conductive carbon material such as graphene was carried out. The performance of material and DSSC was characterized using X-ray diffraction, SEM, FTIR, diffuse reflectance UV-Vis spectroscopy and potentiostat. Studies showed that the incorporation of graphene enhanced the absorption and decreased the band gap resulting in an increase in the electrical conductivity and decrease in charge recombination rate. Moreover, IV measurements of the G-Nb_2O_5 cells showed the 68% increase in fill factor and 52% increase in efficiency (PCE) as compared to cells using Nb_2O_5 nanoparticles as a photoanode.

Effendi et al. [37] fabricated a Gr-ZnO nanocomposite photoanode for DSSCs by electrodeposition process. The DSSCs based on Gr-ZnO nanocomposite were studied through electrochemical impedance spectroscopy (EIS), UV-Visible diffused reflectance spectroscopy (UV-Vis) and photovoltaic performances J-V curves. EIS showed that a lesser charge transport time constant arisen in DSSCs based on Gr-ZnO nanocomposite as compared to ZnO. This enhanced the electron collection efficiency, resulting in high open-circuit voltage. Moreover, Gr-ZnO nanocomposite showed an effective photo-induced charge separation and transportation, thus exhibiting maximum photocurrent response as compared to ZnO. Moreover, The J-V curves demonstrated that incorporating graphene into the ZnO photoanode can permit the DSSC devices operate more efficiently. Gr-ZnO nanocomposite showed a maximum power conversion efficiency of 7.01%, which is doubled from pure ZnO photoanode.

Chou et al. [32] fabricated GO-TiO_2 and zinc oxide nanowires (ZNWs) as photoelectrode layer of DSSC. The morphology and effects of ZNWs/GO-TiO_2 were examined by the field emission scanning electron microscopy (FESEM), UV-visible spectrometer and solar simulator. The improvement of the photovoltaic performances of DSSC was due to the high specific surface area of GO and high electron mobility of ZnO. In addition, the electrochemical properties of GO-TiO_2/ZNWs double structure were studied by electrochemical impedance spectroscopy. The high surface area of GO and the high electron mobility of ZnO could enhance the photovoltaic performances of DSSCs. Compared with the different composite films, the ZNWs/GO-TiO_2 obviously showed higher dye loading. The optimal 3.82% PCE was achieved in DSSCs when the ZNWs/GO-TiO_2 composite film was modified with 1.5 mL GO solution, which was 70.09% more than TiO_2-based DSSC (3.82%).

Batmunkh et al. [38] used SnO_2 and reduced SnO_2-RGO as photoanode in DSSCs synthesized by microwave-assisted method. The incorporation of RGO into SnO_2 resulted in greater electron transfer, thus enhancing the PCE of device by 91.5%. The improvement in efficiency can be ascribed to increase loading of dye, addition of suitable energy levels and enhanced electron transport of the photoanode (**Table 1**).

Bykkam et al. [39] used few-layered graphene (FLG)/SnO_2 nanocomposite as a photoanode for DSSC. FLG varied from 1 to 3 wt %, and their effect was examined on the PCE. XRD patterns confirmed the existence of FLG and SnO_2 nanoparticles in FLG/SnO_2 nanocomposites. The band gap of SnO_2 nanoparticles was found to be ~4.239, 4.237, 4.210 and 4.172 eV for FLG (1.0, 2.0 and 3.0 wt%)/SnO_2 nanocomposites, respectively. An increased 3.02% PCE was observed for 1 wt% FLG/SnO_2 nanocomposite as compared to pure SnO_2 NPs. These results validated that suitable ratio of (1 wt%) FLG in SnO_2 achieves the role of blocking layer to decrease the back electron-hole recombination in DSSC and thus enhance the PCE, though the higher wt% of the FLG causes shielding of the FLG resulting in decrease of efficiency of DSSC.

Photoanode	V_{OC} (V)	J_{SC} (mA/cm^2)	FF (%)	PCE %	Reference
TiO_2	0.71	6.27	60.2	2.68	[10]
TiO_2-2%G			67.1	7.68	[31]
TiO_2-SnO_2-RGO(0.45)	0.67	10.185	46.1	3.16	[38]
FLG 1%-SnO_2	0.54	12.21	45.6	3.02	[39]
G-ZnO/CdS	0.38	8.35	29	0.94	[26]
G-ZnO	0.91	10.89	66	7.01	[37]
G-Nb_2O_5	0.196	0.363	42	0.11	[28]
Ti at 1% GO	0.68	8.42	63.9	3.69	[10]

Table 1.
Photovoltaic parameters of various photoanode materials in DSSCs.

2.2 Graphene application in PSCs

The discovery of organo metal halide perovskite $CH_3NH_3PbX_3$ (X = halogen) as highly efficient light absorber in both photoelectrochemical and photovoltaic cells shed new light on emerging cheaper and highly efficient next-generation solution processed solar cells [7, 40]. Perovskite solar cells are a promising new substitute compared to silicon and dye-synthesized solar cells [2, 8]. PSCs have become the most promising photovoltaic technology, due to their high efficiency (certified 22.1%) and low cost in less than 8 years of development [4, 41, 42]. This extraordinary pace of development has gained much consideration and brought many engineers and scientists to this field.

Perovskite is named after the Russian scientist "Perovski" who first suggested the ABX_3-type (e.g. $CaTiO_3$) crystal structure for perovskites. For solar cell applications, a new class of perovskite material, organo metal trihalide, has been recently introduced. In organo metal trihalide perovskites, "A" is an "organic monovalent cation" (e.g. $CH_3NH_3^+$), "B" is a "metal divalent cation" (e.g. Sn^{2+} or Pb^{2+}) and "X" is a "halide anion" (e.g. Cl^-, Br^- or I^-) [5, 41, 43].

Perovskite materials have exceptional properties such as a suitable band gap, high carrier mobility, broadband absorption of light, long carrier diffusion lengths and ambipolar transport due to which they have been explored as charge-transporting and light-absorbing materials in photovoltaic devices [41, 43]. A big advantage of PSCs is that they can react with light of various wavelengths, converting more sunlight into electricity and resulting in high efficiency. Furthermore, they are lightweight, flexible and semi-transparent. The first perovskite-based dye-sensitized solar cell $CH_3NH_3PbBr_3/TiO_2$ was developed in 2009 which obtained PCE of 3.13%, since then the PSCs have made an outstanding advancement, and in a period of 8 years, the PCE has been improved to around 23% (22.1% certified) [6, 44]. However, PSCs still face challenge of environmental and stability issues, which need to be improved.

Typically, PSCs with regular configuration (**Figure 4**) consists of three main layers:

- Electron transport layer (ETL)
- Perovskite absorber layer
- Hole-transport layer (HTL)

The foremost function of an ETL is the extraction of photo-generated electrons from perovskite and their transference to electrodes. ETL also acts as hole-blocking layer (HBL) [45]. The basic prerequisite for an ideal ETL is high conductivity, excellent electron mobility, high optical transmittance and an appropriate work function [8, 44, 46]. As far as PSCs are concerned, various ETLs such as TiO_2, SnO_2, Nb_2O_5, ZnO, Zn_2SO_4, In_2S_3 and $BaSnO_3$ have been used successfully using different techniques [46].

The HTL lies in the heart of solar cell in between the metal electrode and perovskite. It plays a vital role in the PSC structures and extracts positive charges (holes) from the perovskite and transports them to top electrode. It prevents the direct contact of Perovskite and top electrode [47]. The most widely used hole-transporting material is Spiro-OMeTAD. Other materials used as HTL are PEDOT:PSS, P3HT, PTAA, NiO, CuSCN, etc. [41, 47].

A PSC includes a perovskite (ABX_3)-structured compound as the light-harvesting active layer, most commonly organic–inorganic lead halide-based

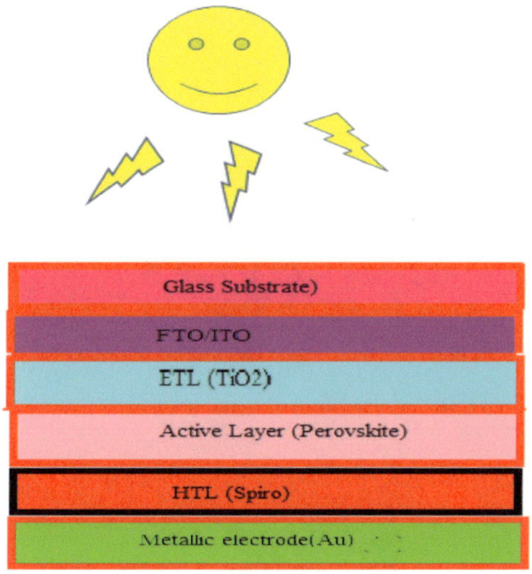

Figure 4.
Representative architecture of PSC.

material. The perovskites commonly used as active layer are methylammonium lead triiodide ($CH_3NH_3PbI_3$) and formamidinium lead triiodide ($CH_3(NH_2)_2PbI_3$) [4, 48]. Perovskite has a cubic crystal (**Figure 5**) structure with three-dimensional (3D) framework and shares BX_6 octahedron with the A ion placed at the octahedral interstices [4, 5, 47, 48]. The perovskite has unique ambipolar properties of generating and transporting both photo-generated holes and electrons. They act both as efficient light absorbers and charge carriers [47].

Theoretically, the working principle of photovoltaic devices involves four general processes as represented by **Figure 6**.

1. Absorption of light

2. Separation of charges

3. Transport of charge

4. Charge collection

Upon light illumination, the perovskite material undergoes photoexcitation and charge separation. The photoexcitation generates an electron-hole pair in the perovskite material. Then, diffusion of electrons and holes occurs toward the interfaces of perovskites/charge selective layers, respectively. After charge carrier (electrons and holes) extractions to charge selective layers, electrons and holes diffuse in the ETM and HTM, respectively, towards the corresponding electrodes for charge collection. These charge carriers produce a potential difference at the back and front contacts of the PSCs and can generate an electric current [4, 41].

In 2009, $MAPbI_3$ was first used as sensitizer by Miyasaka and coworkers in liquid-state DSSCs with a PCE of 3.81% [49]. In 2011, a PCE of 6.5% was reported for perovskite-sensitized DSSCs in which perovskite quantum dots were used as sensitizers [50]. However, a foremost advancement was achieved in 2012, using organometal halide perovskite sensitizers in perovskite-based solid state DSSCs (or PSCs), and PCEs of 9.7 and 10.9% were obtained [4]. Later, more exciting

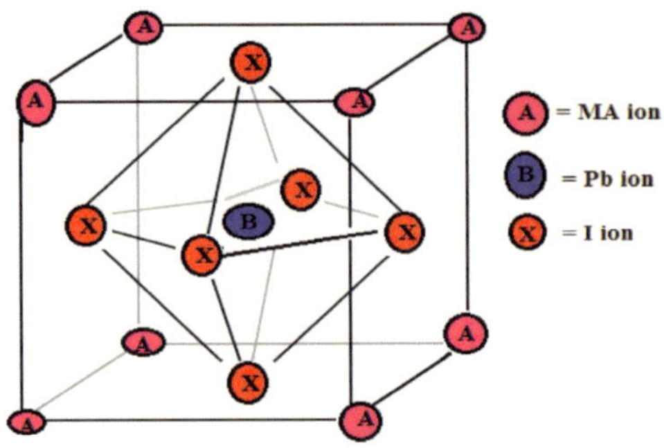

Figure 5.
Crystal structure of $CH_3NH_3PbI_3$ perovskite.

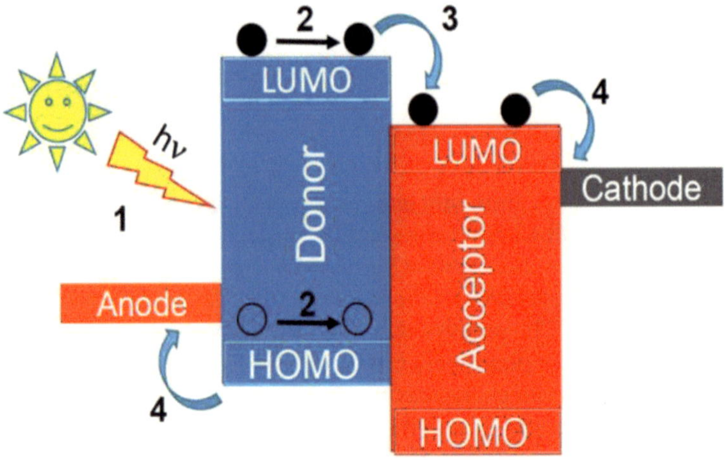

Figure 6.
Working principle of PSC [9].

discoveries and advancement on PSCs arose within a short period of time from 2012 to 2015, and a PCE of 22.1% was achieved [4].

2.2.1 Graphene/metal oxide as electron-transporting layer in PSCs

Though the highest efficiency in solution-processable PSCs have been achieved using metal oxide electron transporting layer, those electrodes require sintering at higher temperature of 500°C, which is not economical and unfavorable for the fabrication of PSCs on plastic substrates. Furthermore, the charge carrier recombination at a metal oxide/perovskite interface is another critical factor in PSCs that delays the charge transportation and decreases the PCE of a device. To report this problem, the graphene and its derivatives and graphene/metal oxide nanocomposites have been as ETL [51]. Recently, some graphene/metal oxide nanocomposites such as G/ZnO, G/TiO_2, G/SnO_2, etc. have been used successfully as photoanode in PSCs.

Graphene nanocomposite with TiO_2 as a low-cost, low-temperature solution processed collection layer in mesostructured PSCs was fabricated by Wang et al. [52]. Graphene nanoflakes provided greater charge collection in nanocomposites, allowing the fabrication of devices at lower temperatures (150°C). These devices showed extraordinary photovoltaic performance with maximum PCE up to 15.6%. Hence, G/TiO_2 nanocomposites demonstrated its possibility to contribute significantly towards the progress of low-cost solar cells.

Han et al. [53], reported reduced graphene oxide/mesoporous (mp)-TiO_2 nanocomposite-based mesostructured PSCs that demonstrated an improved electron transport property due to the reduced interfacial resistance. The amount of rGO added to the TiO_2 nanoparticles was optimized, and their effects on electron diffusion, film resistivity, recombination time and photovoltaic performance were investigated. The rGO/mp-TiO_2 film decreases interfacial resistance when compared to the mp-TiO_2 film, and, thus, it improves charge collection efficiency. This effect significantly increases the VOC and ISC. The rGO/mp-TiO_2 nanocomposite film with an optimal rGO 0.4 vol% content showed 18% higher PCE than the PSCs using TiO_2 nanoparticles.

Saleem et al. [33] reported a new, low-temperature solution-processing approach to use rGO-TiO_2 composite material for the electron transport layer of PSCs. GO was synthesized by a modified Hummer's method and TiO_2 nanoparticles by hydrothermal method. Thin films of GO-TiO_2 were made with different wt.% of GO on indium tin oxide (ITO) substrate by spin coating, followed by annealing at 150°C. The band gap of the pure TiO_2 thin film was calculated to be 3.5 eV, which was reduced to 2.9 eV for the GO-TiO_2 nanocomposites with a red shift towards higher wavelength. Moreover, thermal post annealing at 400°C enhanced the transparency in the visible region and decreased the sheet resistance. The I-V study indicated an ohmic contact with the ITO substrate.

Chandrasekhar and Komarala [51] fabricated PSCs using a graphene/ZnO nanocomposite (G/ZnO NC) as an electron-transporting layer. ZnO and G/ZnO NC films were developed by a novel spray deposition method compatible with large-area processing methods for deposition of pristine. The effect of amount of graphene concentration was studied, and GO amount varied from 0 to 1 wt% in the G/ZnO NC films. It was found that a 0.75 wt% GO concentration in the G/ZnO NC films gives the best PSC performance with ISC and PCE going up from 15.54 to 19.97 mA cm^{-2} and 7.01 to 10.34%, respectively, as compared to pristine ZnO. The improvement in PV performance is ascribed to the higher growth of the perovskite thin film and enhanced electron transport/extraction by using the graphene amount in the NC.

Xie et al. [54] improved the electronic properties of SnO_2 by adding minute amount of graphene quantum dots (GQDs) and reported improvement in electronic properties of SnO_2. The photo-generated electrons in GQDs can be easily

Electron-transporting layer	VOC (V)	J_{SC} (mA/Cm^2)	FF (%)	PCE (%)	Reference
TiO_2 NP	1.0	17.7	61.0	10.1	[52]
High-temperature-sintered TiO_2	1.00	21.4	70.0	14.1	[52]
SnO_2	1.10	22.1	73.6	17.91	[54]
SnO_2-sintered free	0.91	20.73	64,2	12.10	[42]
SnO_2 nanobelts	1.08	22.46	66.0	16.02	[46]
SnO_2/GQDs	1.134	23.05	77.8	20.31	[54]
G-SnO_2	1.84	22.66	82.1	20.16	[55]
(0.4 vol %) rGO/mp-TiO_2	0.91	21.0	70.8	13.5	[53]
TiO_2 + graphene	1.04	21.9	73.1	15.6	[52]
0.75 wt% G/ZnO	0.926	19.97	56.3	10.34	[51]

Table 2.
Photovoltaic parameters of various electron-transporting materials in PSCs.

transferred to the conduction band of SnO_2, thus enhancing the conductivity of SnO_2 and reducing the charge recombination at the interface. SnO_2/GQDs fabricated solar cell showed low hysteresis and average PCE of 19.2%.

Xiaojuan et al. [55] tried to incorporate chemical modified (2D naphthalene diimide) graphene into SnO_2 nanocrystal as ETL for highly efficient PSCs. They modified SnO_2 with 2D naphthalene diimide-graphene, which can increase the surface hydrophobicity and is responsible for the van der Waals interactions between surfactant and perovskite. Thus, highly efficient PSCs were fabricated with maximum PCE of 20.2% and enhanced fill factor of 82%, which could be attributed to the improved electron extraction ability, electron mobility and the reduced carrier recombination, resulting in the increased FF. This work provides an important direction for further search in utilizing carbonaceous materials for low-temperature solution-processed planar PSCs (**Table 2**).

3. Conclusions

Graphene, due to its unique properties, has been explored widely in solar cells, and graphene metal oxide nanocomposites have been widely used in emerging third-generation solar cells. In view of reported literature, it is obvious that these materials played a significant role in enhancing the efficiency of 3G solar cells. In this chapter, the use of graphene nanocomposites has been explored in DSSCs and PSCs. Graphene nanocomposites as photoanode material in DSSCs seemed to be a promising approach to increase the charge transport, charge separation and thus the performance of solar cell. The low cost of graphene brings new prospects for commercialization, particularly to replace the use of Pt as counter electrode in DSSCs. In PSCs, the use of graphene and its derivatives with metal oxide (G/metal oxide nanocomposites) as electron transport layer has shown to be valuable. Incorporating of graphene into metal oxide reduces the series resistance and the charge recombination between perovskite and metal oxide as well as improves the electron and hole transport from the perovskite layer to the corresponding electrodes. Moreover, graphene nanocomposites have proved to be an efficient electron-transporting layer fabricated at lower temperature, thus suitable for flexible substrates in PSCs. Furthermore, these materials are interesting to replace Pt-FTO or ITO and noble metals (silver or gold) in 3G solar cells. Thus, the efficient replacement of the existing expensive materials reduces the cost of third-generation solar cells.

4. Futuristic aspects/prospective of graphene metal oxide nanocomposites

Graphene-based materials are interesting from the perspective of both fundamental science and technology due to their nontoxicity, chemical and thermal stability and mechanical strength. Although graphene metal oxide nanocomposites can be used to enhance the PCE of 3G solar cells by various optimization methods, manufacturing of highly efficient and stable solar cell is challenging. Moreover, much practical work needs to be done to obtain the optimal performance. From the future perspective, a mechanism for physical-chemical interactions of these nanocomposites in different layers of solar cells is necessary.

From energy point of view, graphene metal oxide nanocomposites can be used for energy storage and conversion devices such as solar cells, fuel cells, batteries, capacitors, etc. The unique graphene properties, such as high conductivity and transparency, offer to replace the ITO, which lacks the flexibility and robustness in flexible solar cells, light-emitting diodes (LEDs), touch screens and displays. Graphene/metal oxide nanocomposites can be used to improve the durability of potential optoelectronic devices. Graphene/metal oxide nanocomposites are promising substitutes to decrease the drawbacks of using only metal oxide nanoparticles in various applications, such as anode materials in lithium ion batteries (LIBs), sensors, photocatalysts, removal of organic pollutants, etc. The graphene/metal oxide nanocomposites showed greater versatility as enhanced materials for the fabrication of electrochemical sensors and biosensors.

Graphene, due to its antibacterial, antiplatelet and anticancer activities, make it a potential candidate for biological and biomedical applications, and graphene/metal oxide nanocomposites can be used in cancer therapy, drug delivery, bioimaging, tissue engineering, etc. The synthesis, toxicity, biocompatibility and biomedical applications of graphene/metal oxide nanocomposites are important issues that require thorough investigation in any kind of applications related to human welfare.

Despite the substantial progress in the synthesis of graphene-based nanocomposites, challenges exist in the application on an industrial scale. For example, advanced applications of graphene/metal oxide nanocomposites require wide research to understand the interactions between graphene surface and the nanomaterials, which will have direct influence on the properties of these nanocomposites. An appropriate understanding of these interactions will surely enhance the application potential of the nanocomposites in various fields, such as catalysis, biosensing, drug delivery, imaging etc. Therefore, graphene metal oxide composites have vast potential for many industrial applications, and they are commercially feasible compared to carbonaceous-based nanocomposites.

Author details

Tahira Mahmood*, Madeeha Aslam and Abdul Naeem
National Centre of Excellence in Physical Chemistry, University of Peshawar, Peshawar, Pakistan

*Address all correspondence to: tahiramahmood@uop.edu.pk

References

[1] Liu X, Fang J, Liu Y, Lin T. Progress in nanostructured photoanodes for dye-sensitized solar cells. Frontiers of Materials Science. 2016;**10**(3):225-237. DOI: 10.1007/s11706-016-0341-0

[2] Choi JW, Jin JW, Tondelier D, Bonnassieux Y, Geffroy B. Low temperature solution-processable 3D-patterned charge recombination layer for organic tandem solar cells. Materials. 2019;**12**. DOI: 10.3390/ma12010162

[3] Bagher AM, Vahid MMA, Mohsen M. Types of solar cells and application. American Journal of Optics and Photonics. 2015;**3**:94-113. DOI:10.11648/j.ajop.20150305.17

[4] Yun S, Qin Y, Uhl AR, Nikolaos V, Yin M, Li D, et al. New-generation integrated devices based on dyesensitized and perovskite solar cells. Energy & Environmental Science. 2018. DOI: 10.1039/C7EE03165C

[5] Zhou D, Zhou T, Tian Y, Zhu X, Tu Y. Perovskite-based solar cells: Materials, methods, and future perspectives. Journal of Nanomaterials. 2018;**2018**. DOI: 10.1155/2018/8148072

[6] Willian DSLF, Andriy Z. Research update: Emerging chalcostibite absorbers for thin-film solar cells. APL Materials. 2018;**6**:084501-084513. DOI: 10.1063/1.5027862

[7] Yang S, Fu W, Zhang Z, Chen H, Li C-Z. Recent advances in perovskite solar cells: Efficiency, stability and leadfree perovskite. Journal of Materials Chemistry A. 2017. DOI: 10.1039/C7TA00366H

[8] Thakur UK, Kisslinger R, Shankar K. One-dimensional electron transport layers for perovskite solar cells. Nanomaterials. 2017;7. DOI: 10.3390/nano7050095

[9] Ahmadpour M. Metal-oxide based interlayers for organic and perovskite photovoltaics [Thesis]. University of Southern Denmark; 2017

[10] Kanta U-A, Thongpool V, Sangkhun W, Wongyao N, Wootthikanokkhan J. Preparations, characterizations, and a comparative study on photovoltaic performance of two different types of graphene/TiO_2 nanocomposites photoelectrodes. Journal of Nanomaterials. 2017;**2017**. DOI: 10.1155/2017/2758294

[11] Jana A, Scheer E, Polarz S. Synthesis of graphene–transition metal oxide hybrid nanoparticles and their application in various fields. Beilstein Journal of Nanotechnology. 2017;**8**: 688-714. DOI: 10.3762/bjnano.8.74

[12] Patil K, Rashidi S, Wang H, Wei W. Recent progress of graphenebased photoelectrode materials for dye-sensitized solar cells. International Journal of Photoenergy. 2019;**2019**:1-16. DOI: 10.1155/2019/1812879

[13] Haldorai Y, Yeon S-H, Huh YS, Han Y-K. Electrochemical determination of tryptophan using a glassy carbon electrode modified with flower-like structured nanocomposite consisting of reduced graphene oxide and SnO_2. Sensors and Actuators B. 2017;**239**: 1221-1230. DOI: 10.1016/j.snb.2016.09.119

[14] Ribeiro-Soares J, Dresselhaus MS. News and views: Perspectives on graphene and other 2D materials research and technology investments. Brazilian Journal of Physics. 2013. DOI: 10.1007/s13538-013-0156-8

[15] Chowdhury TH, Islam A, Hasan AKM, Terdi MA, Singh SP, Alam MK, et al. Prospects of graphene as a potential carrier-transport material in third-generation solar cells. Chemical

Record. 2016;**16**:614-632. DOI: 10.1002/tcr.201500206

[16] Chen D, Feng H, Li J. Graphene oxide: Preparation, functionalization, and electrochemical applications. Chemical Reviews. 2010. DOI: 10.1021/cr300115g

[17] Bonaccorso F, Colombo L, Yu G, Stoller M, Tozzini V, Ferrari AC, et al. Graphene, related two-dimensional crystals, and hybrid systems for energy conversion and storage. Science. 2015;**34**. DOI: 10.1126/science.1246501

[18] Ciesielski A, Samorí P. Graphene via sonication assisted liquid-phase exfoliation. Chemical Society Reviews. 2013. DOI: 10.1039/c3cs60217f

[19] Iwan A, Chuchmała A. Perspectives of applied graphene: Polymer solar cells. Progress in Polymer Science. 2012;**37**:1805-1828. DOI: 10.1016/j.progpolymsci.2012.08.001

[20] Grande L, Chundi VT, Wei D, Bower C, Andrew P, Ryhänen T. Graphene for energy harvesting/storage devices and printed electronics. Particuology. 2012;**10**:1-8. DOI: 10.1016/j.partic.2011.12.001

[21] Gurunathan S, Kim J-H. Synthesis, toxicity, biocompatibility, and biomedical applications of graphene and graphene-related materials. International Journal of Nanomedicine. 2016;**11**:1927-1945. DOI: 10.2147/IJN.S105264

[22] Xu G, Malmström J, Edmonds N, Broderick N, Travas-Sejdic J, Jin J. Investigation of the reduction of graphene oxide by lithium triethylborohydride. Journal of Nanomaterials. 2016;**2016**:1-10. DOI:10.1155/2016/4021059

[23] Yoo MJ, Kim HW, Yoo BM, Park HB. Highly soluble polyetheramine functionalized graphene oxide and reduced graphene oxide both in aqueous and non aqueous solvents. Carbon. 2014;**75**:149-160. DOI: 10.1016/j.carbon.2014.03.048

[24] Hu C, Lu T, Chen F, Zhang R. A brief review of graphene–metal oxide composites synthesis and applications in photocatalysis. Journal of The Chinese Chemical Society. 2013;**1**:21-39. DOI: 10.1080/22243682.2013.771917

[25] Szostak R, Morais A, Carminati SA, Costa SV, Marchezi PE, Nogueira AF. Application of graphene and graphene derivatives/oxide nanomaterials for solar cells. In: The Future of Semiconductor Oxides in Next- Generation Solar Cells. Elsevier; 2018. pp. 398-437. DOI: 10.1016/ B978-0-12-811165-9.00010-7

[26] Ghoreishi FS, Ahmadi V, Samadpour M. Synthesis and characterization of graphene-ZnO nanocomposite and its application in photovoltaic cells. Journal of Nanostructures. 2013;**3**:453-459

[27] Hosen MM. Investigation of chemical and electrochemical behaviour of graphene/manganese oxide/tin oxide ternary nanocomposite [Thesis]. Bangladesh: University of Engineering and Technology; 2017

[28] Jamil M, Khan ZS, Ali A, Iqbal N. Studies on solution processed graphene-Nb_2O_5 nanocomposite based photoanode for dye-sensitized solar cells. Journal of Alloys and Compounds. 2017;**694**:401-407. DOI: 10.1016/j.jallcom.2016.09.300

[29] Eshaghi A, Aghaei AA. Effect of TiO_2–graphene nanocomposite photoanode on dye-sensitized solar cell performance. Bulletin of Materials Science. 2015;**38**:1177-1182

[30] Nouri E, Mohammadi MR, Lianos P. Impact of preparation method of TiO_2-RGO nanocomposite photoanodes on the performance of dyesensitized solar cells.

Electrochimica Acta. 2016;**219**:38-48. DOI: 10.1016/j.electacta.2016.09.150

[31] Kazmi SA, Hameed S, Ahmed AS, Arshad M, Azam A. Electrical and optical properties of graphene-TiO_2 nanocomposite and its applications in dye sensitized solar cells (DSSC). Journal of Alloys and Compounds. 2017;**691**:659-665. DOI: 10.1016/j.jallcom.2016.08.319

[32] Chou J-C, Member I, You P-H, Liao Y-H, Lai C-H, Chu C-M, et al. Fabrication and photovoltaic properties of dye-sensitized solar cells based on graphene–TiO_2 composite photoelectrode with ZnO nanowires. IEEE Transactions on Semiconductor Manufacturing. 2017;**30**:531-538. DOI: 10.1109/TSM.2017.2747121

[33] Saleem A, Ullah N, Khursheed K, Iqbal T, SHAH SA, Asjad M, et al. Graphene oxide TiO_2 nanocomposite for electron transport applications. Journal of Electronic Materials. 2018;**47**:3749-3756. DOI: 10.1007/s11664-018-6235-4

[34] Dinari M, Momeni MM, Goudarzirad M. Dye-sensitized solar cells based on nanocomposite of polyaniline/graphene quantum dots. Journal of Materials Science. 2016;**51**:2964-2971. DOI: 10.1007/s10853-015-9605-9

[35] Wang X, Zhi L, Mullen K. Transparent, conductive graphene electrodes for dye-sensitized solar cells. Nano Letters. 2008;**8**:323-327

[36] Jang S-R, Vittal R, Kim K-J. Incorporation of functionalized singlewall carbon nanotubes in dye-sensitized TiO_2 solar cells. Langmuir. 2004;**20**:9807-9810. DOI: 10.1021/la049022f

[37] Effendi NAS, Samsi NS, Zawawi SA, Hassan OH, Zakaria R, Yahya MZA, et al, editors. Studies on graphene zinc-oxide nanocomposites Photoanodes for high-efficient dyesensitized solar cells. In: AIP Conf Proc Adv Mater Nanotechnol (ICAMN IV 2016). AIP; 2017;**2017**

[38] Batmunkh M, Dadkhah M, Shearer CJ, Biggs MJ, Shapter JG. Incorporation of graphene into SnO_2 photoanode for dye-sensitized solar cells. Applied Surface Science. 2016;**387**:690-697. DOI: 10.1016/j.apsusc.2016.06.146

[39] Bykkam S, Rao KV, Chakra CS. Enhanced photovoltaic efficiency of dye sensitized solar cell by few layered graphene sheet decked with SnO_2 nanopartilces as a photoanode. Journal of Nanoscience and Nanotechnology. 2016;**2**:144-147

[40] Greul E, Docampo P, Bein T. Synthesis of hybrid tin halide perovskite solar cells with less hazardous solvents: Methanol and 1,4-dioxane. Journal of Inorganic and General Chemistry. 2017;**643**:1704-1711. DOI: 10.1002/zaac.201700297

[41] Chen Y, Zhang L, Zhang Y, Gao H, Yan H. Large-area perovskite solar cells – A review of recent progress and issues. RSC Advances. 2018;**8**:10489-10508. DOI: 10.1039/C8RA00384J

[42] Zhao J, Wei L, Liu J, Wang P, Liu Z, Jia C, et al. A sintering-free, nanocrystalline tin oxide electron selective layer for organometal perovskite solar cells. Science China Materials. 2017;**60**(3):208-216. DOI: 10.1007/s40843-016-5159-x

[43] Shi Z, Jayatissa AH. Perovskitesbased solar cells: A review of recent progress, materials and processing methods. Materials (Basel). 2018;**11**(5):1-34. DOI: 10.3390/ma11050729

[44] Uddin A, Upama M, Yi H, Duan L. Encapsulation of organic

and perovskite solar cells: A review. Coatings. 2019;**9**(2). DOI: 10.3390/coatings9020065

[45] Jiang Q, Zhang L, Wang H, Yang X, Meng J, Liu H, et al. Enhanced electron extraction using SnO_2 highefficiency planar-structure $HC(NH_2)_2PbI_3$ for -based perovskite solar cells. Nature Energy. 2016. DOI: 10.1038/nenergy.2016.177

[46] Mali SS, Patil JV, Kim H, Hong CK. Synthesis of SnO_2 nanofibers and nanobelts electron transporting layer for efficient perovskite solar cells. Nanoscale. 2018. DOI: 10.1039/C8NR00695D

[47] Vivo P, Salunke JK, Priimagi A. Hole-transporting materials for printable perovskite solar cells. Materials. 2017;**10**:1087. DOI: 10.3390/ma10091087

[48] Green MA, Ho-Baillie A, Snaith HJ. The emergence of perovskite solar cells. Nature Photonics. 2014;**8**. DOI: 10.1038/nphoton.2014.134

[49] Kojima A, Teshima K, Shirai Y, Miyasaka T. Organometal halide perovskites as visible-light sensitizers for photovoltaic cells. Journal of the American Chemical Society. 2009;**131**:6050-6051. DOI: 10.1021/ja809598r

[50] Im JH, Lee CR, Lee JW, Park SW, Park NG. 6.5% efficient perovskite quantum-dot-sensitized solar cell. Nanoscale. 2011;**3**(10):4088-4093. DOI: 10.1039/c1nr10867k

[51] Chandrasekhar PS, Komarala VK. Graphene/ZnO nanocomposite as an electron transport layer for perovskite solar cells; the effect of graphene concentration on photovoltaic performance. RSC Advances. 2017;7:28610-28615. DOI: 10.1039/C7RA02036H

[52] Wang JT-W, Ball JM, Barea EM, Abate A, Alexander-Webber JA, Huang J, et al. Low-temperature processed Electron collection layers of graphene TiO_2 in thin film perovskite solar cells. Nano Letters. 2014;**14**:724-730. DOI: 10.1021/nl403997a

[53] Han GS, Song YH, Jin YU, Lee J-W, Park N-G, Kang BK, et al. Reduced graphene oxide/mesoporous TiO_2 nanocomposite based perovskite solar cells. ACS Applied Materials & Interfaces. 2015;7:23521-23526. DOI: 10.1021/acsami.5b06171

[54] Xie J, Huang K, Yu X, Yang Z, Xiao K, Qiang Y, et al. Enhanced electronic properties of SnO_2 via electron transfer from graphene quantum dots for efficient perovskite solar cells. ACS Nano. 2017;**11**:9176-9182. DOI: 10.1021/acsnano.7b04070

[55] Xiaojuan Z, Tao L, Li H, Huang W, Sun P, Liu J, et al. Efficient planar perovskite solar cells with improved fill factor via Interface engineering with graphene. Nano Letters. 2018. DOI: 10.1021/acs.nanolett.8b00025

Section 3

3-D Metal - Non-Metal Matrix

Chapter 3

Titanium Dioxide Versatile Solid Crystalline: An Overview

Lourduraj Stephen

Abstract

Among the several choices, titanium dioxide (TiO_2) is the most efficient material and has attracted great attention because of its certain specific properties like high permittivity, refractive index, efficiency, low cost, chemical inertness, non-toxicity, photocatalytic activity, photostability and capability of decomposing a wide variety of organic compounds. In the field of dental, orthopedic and osteosynthesis applications, the titanium and its native oxide (titanium dioxide) are used as an implant material. TiO_2 is used in an extremely wide range of commercial applications and research areas including: (i) TiO_2 powder: as a white pigment in paint, plastic, inks, paper and cosmetics; in washing powder, toothpaste, sunscreen, foodstuffs, pharmaceuticals, photographic plates, for creating synthetic gemstones; and as a catalyst. (ii) TiO_2 thin films: for ultra-thin capacitors and MOSFETs due to its extremely high dielectric constant; as humidity and oxygen sensor due to the dependence of its electrical conductance on the gases present; as an optical coating and a material for waveguides due to its high refractive index; as a protective coating and corrosion-resistant barrier; and as a photoanode in solar cells due its photoelectric activity.

Keywords: nanoparticles, titania, sol-gel, spin coating, thin films, electronic, optical, solar cells

1. Introduction

Research in the development of efficient materials has seen significant progress in the last two decades with a large number of research works carried out every year. Improvements in the performance of materials have been largely correlated with advances in nanotechnology. In recent years, the metal oxide nanoparticles are increasingly receiving attention for their wide range of applications in almost each and every field. Concerns regarding metal oxide nanoparticles exist in their chemistry and size and for being non-biodegradable. This poses the rapid distribution of nanoparticles in the environment with potentially harmful consequences.

2. Titanium dioxide semiconducting material

Titanium dioxide (TiO_2) is an n-type metal oxide semiconducting material used in a wide range of common and high-tech applications. It is cheap, chemically stable, non-toxic and bio-compatible. Titania is successfully used as implant material in dental, orthopedic and osteosynthesis applications and its native oxide mostly constitutes titanium dioxide [1]. TiO_2 in the form of nanopowder is used as a white pigment in paint [2], replacing lead oxide that is toxic, and in toothpaste. In the form of solid thin films, transparent single crystals or its thin films have a high refractive index that makes TiO_2 suitable for optical applications [3, 4]. Multi-layers composed of TiO_2 and SiO_2 are designed to make antireflection coatings in the whole visible range [5, 6]. TiO_2 is widely used for photocatalysis [6]; for example, electrodes made of TiO_2 are used in electrochromic devices [7] and dye-sensitized solar cells [8] etc., and solid-state photovoltaic solar cells with porous TiO_2 layer show promising results [2, 3]. Pd-TiO_2 diodes are used as hydrogen gas sensors [9, 10], and nowadays, TiO_2 can replace ZrO_2 in lambda probes used in the car industry [11]. Most of these applications are due to its n-type semiconducting property and realized with micro- or nano-structured TiO_2 nanopowders or nano thin films.

3. TiO_2 crystal structure

Titanium oxide exists in nature as minerals and it has various structures under ambient conditions: rutile, anatase, brookite and srilankite (this last structure is also called PbO_2-type TiO_2 or TiO_2-II) [12–14]. Rutile is a relatively abundant material and its structure is the most stable [13] and also the most studied. Anatase and brookite are extremely rare in nature [10]. TiO_2 thin films are generally amorphous for deposition temperatures ≤350°C, above which anatase is formed. The most stable crystalline phase, rutile, is formed at temperatures greater than about 800°C. The brookite phase is rarely observed in deposited thin films. The functional properties of TiO_2 films, powders and ceramics are strongly dependent on the phase of the material. Thus, TiO_2-x is an n-type semiconductor, in contrast with p-type semiconductors, which contain electron acceptors and where the charge carriers are holes rather than electrons [15]. Substoichiometric TiO_2-x is both a poor insulator and a modest semiconductor. Therefore, several attempts have been made either to control the oxygen vacancy concentration or to introduce charge carriers (doping) inside TiO_2 in order to enhance the properties, depending on the needed application. After 50 years, the periodic table was updated by incorporation of new atoms such as Ti and so on [16]. The structural, optical and electrical properties of TiO_2 are reported in **Table 1**.

Polymorph	Anatase	Rutile	Brookite
Structure and space group [1]	Tetragonal I 4_1/amd	Tetragonal P 4_2/mnm	Orthorhombic Pbca
Lattice parameter	a = 3.7923 c = 9.5548 (T = 350°C)	a = 4.5930 c = 2.9590	a = 5.4558 b = 9.1819 c = 5.142
Density (g cm^{-3}) [1]	3.84	4.26	4.12
Refractive index λ = 600 nm [7]	⊥ to c axis 2.55 // to c axis 2.48	⊥ to c axis 2.60 // to c axis 2.89	⊥ to a or b axis 2.57 // to c axis 2.69
Dielectric constant [8–10]	⊥ to c axis 31 // to c axis 48	⊥ to c axis 89 // to c axis 173	78
Band gap (eV) [12, 13]	⊥ to c axis Direct 3.42 // to c axis indirect 3.46	⊥ to c axis Direct 3.04 // to c axis indirect 3.05	3.14
Electron mobility (10^{-4} m^2/Vs) [14, 15]	crystal: 15–550 thin film: 0.1–4	crystal: 0.1–10 thin film: 0.1	—

Table 1.
TiO_2 properties.

The elementary cells of the TiO_2 crystal structures, phase transition and crystallographic structures of TiO_2 are presented in **Figures 1–3**, respectively. Rutile and anatase, which are tetragonal, are more ordered than the orthorhombic structure. Anatase, which is the least dense structure, has empty channels along the a and b axes.

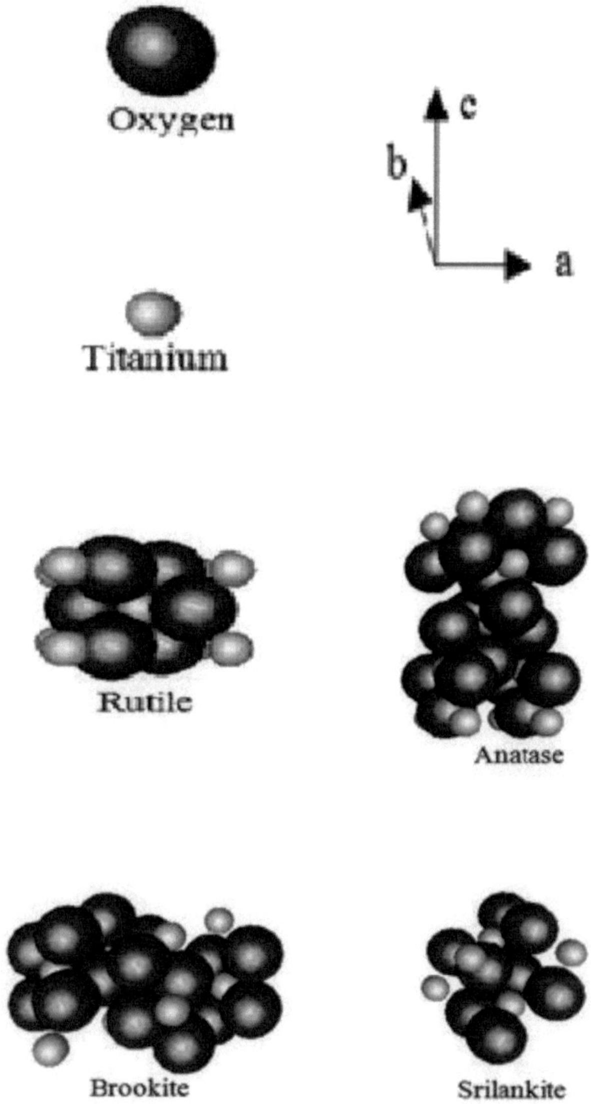

Figure 1.
Elementary cell of TiO_2 polymorphs.

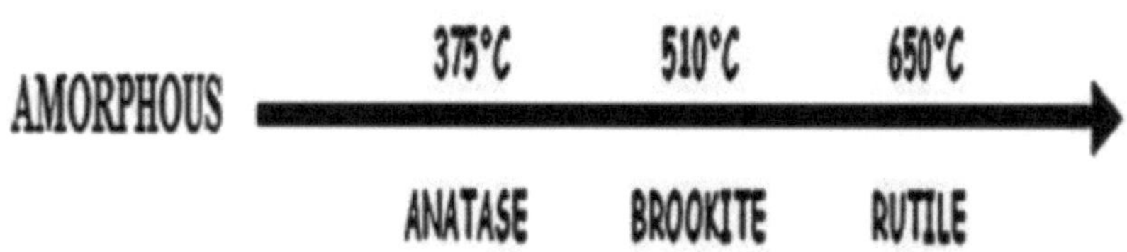

Figure 2.
Phase transition of titanium dioxide.

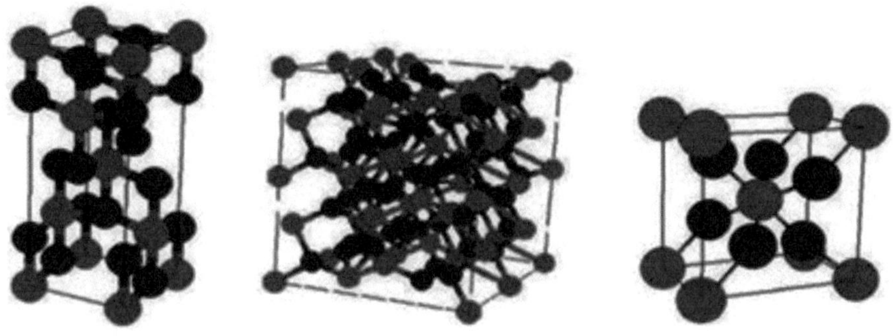

Figure 3.
Crystallographic structures of TiO_2 (a) anatase, (b) brookite, and (c) rutile.

4. Properties of TiO_2 thin films

The performance of TiO_2 thin-film based devices depends on its structural, surface morphological, compositional, optical and electrical properties. It is evident that the improvement of materials properties requires a closer inspection of preparative conditions and also the above said properties of the films.

The physical, optical, electrical and chemical properties of titanium dioxide (TiO_2) depend greatly on the amorphous or crystalline phase of the material. TiO_2 is a complex material with three crystalline phases, two of which are commonly observed in thin films-anatase and rutile. Anatase is commonly observed at film deposition temperatures of 350–700°C, while higher temperatures promote the growth of rutile. Deposition temperature lower than 300°C generally result in the formation of amorphous TiO_2 and it has highest band gap (3.5 eV), low refractive index (1.9–2 at 600 nm) and extinction coefficient. Polycrystalline anatase thin films with an optical band gap (3.2 eV) exhibit a much higher refractive index and slightly increased absorption coefficient. The chemical resistance of amorphous TiO_2 films is poor in many acidic and basic solutions as compared with crystalline structure because of anatase, which is insoluble in many acids and base. The TiO_2 thin films with rutile phase are having extremely higher refractive indices (up to 2.7 at 600 nm) and lower the band gap absorption is still low. The chemical resistance of rutile is excellent, and after annealing at temperatures above 1000°C, it is insoluble in nearly all acids and bases.

Thin films of TiO_2 are used in an extremely wide range of commercial applications and research areas, including the following:

TiO_2 powders and nanopowders: as a white pigment in paint, plastic, inks, paper and cosmetics; in washing powder, toothpaste, sunscreen, foodstuffs, pharmaceuticals, photographic plates, for creating synthetic gemstones; and as a catalyst.

TiO_2 thin films and their derivatives: for ultra-thin capacitors and MOSFETs due to their extremely high dielectric constant; as humidity and oxygen sensor due to the dependence of their electrical conductance on the gases present; as an optical coating and a material for waveguides due to their high refractive index; as a protective coating and corrosion resistant barrier; and as a photoanode in solar cells due their photoelectric activity.

4.1 Semiconductor properties

Solid materials are classified in three groups depending on their electrical conductivity σ. Highly conducting materials are metals ($\sigma > 10^4\ S\ m^{-1}$), material with very low electrical conductivity are insulators ($\sigma < 10^{-8}$ S m^{-1}), and in-between stand the semiconductors. The main difference between metal and semiconductor is the fact that for metals, the electrical conductivity decreases when temperature increases, while the reverse phenomenon usually occurs in the case of semiconductors. The energy band diagram of a pure semiconductor containing a negligible amount of impurities (intrinsic semiconductor) is characterized by an energy gap (*EG*) inside which no electronic states are encountered.

When a semiconductor is doped with donor and/or acceptor impurities, impurity energy levels are introduced. A donor level is defined as being neutral if filled with an electron and positive if empty. An acceptor level is neutral if empty and negative if filled by an electron. The Fermi level for the intrinsic semiconductor lies close to the middle of the band gap (**Figure 4**). When impurity atoms are introduced, the Fermi level must adjust itself to preserve charge neutrality, and the total negative charge (electrons and ionized acceptors) must equal the total positive charge (holes and ionized donors). N-type and p-type semiconductor band diagram is shown in **Figure 4** [17, 18].

4.2 Physical properties

4.2.1 The amorphous-anatase-rutile phase transformations

Amorphous TiO_2 thin films can be deposited at temperatures as low as 100–150°C [19, 20]. Amorphous TiO_2 does not have a strict crystallographic structure, often incorporates voids within the material, and has a relatively low density. For TiO_2 thin films formed by chemical reaction, the lowest temperature crystalline phase of TiO_2 that can be obtained is anatase. To obtain polycrystalline anatase, the film can be either deposited as amorphous TiO_2 and then crystallized by annealing at a higher temperature or deposited as polycrystalline material directly. The results indicate that the transition from an amorphous to anatase film occurs at about 300–365°C, regardless of whether this is the deposition or annealing temperature. Rutile films are initially observed on silicon substrates at deposition temperatures above 700°C, and more typically from 900 to 1100°C. It should be noted that anatase is a metastable phase of TiO_2, and the

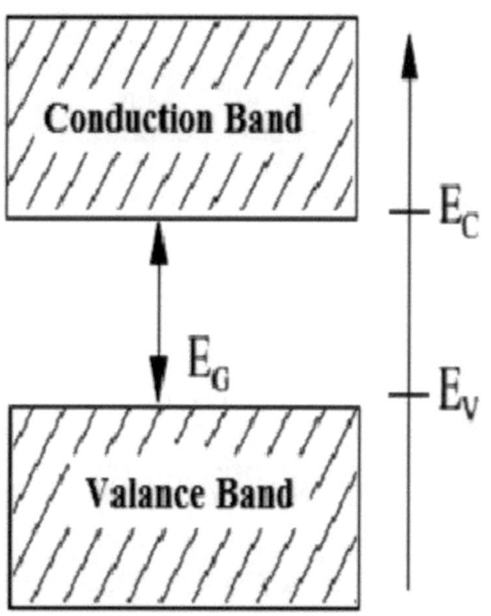

Figure 4.
The band diagram of a semiconductor.

conversion to rutile involves a collapse of the anatase structure, which is irreversible [21, 22]. Although rutile and anatase are both of tetragonal crystallographic structure, rutile is more densely packed and thus possesses a greater density.

The deposition of TiO_2 thin films is formed by chemical reaction, using chemical vapor deposition (CVD), and spray pyrolysis and hydrolysis systems. In this scenario, the substrate temperature is the primary means of controlling the deposited phase of the material. In contrast, physical vapor deposition (PVD) systems, such as evaporation, sputtering, and ion-beam deposition, are used to determine the structure with its phase primarily by the kinetic energy of the impinging atoms.

Therefore, the progression through the amorphous, anatase and rutile phases may not necessarily be expected. This is confirmed by the occurrence of rutile films at low deposition temperatures (< 450°C) by carefully optimized deposition methods, [23, 24] ion-assisted deposition [25] and reactive evaporation [26]. The TiO_2 films are formed by a chemical reaction, where the substrate temperature dominates film growth characteristics. Several researchers observed that the processing temperatures required to convert an anatase film into a rutile one are much higher than temperature required depositing a rutile film directly [26–28]. The variation in physical and chemical properties of the films is determined solely by the maximum processing temperature, whether the deposition temperature or a subsequent annealing temperature was observed by researchers [20].

The mechanism for the sintering and transformation of anatase into rutile involves several steps. Initially, the smallest particles coalesce, forming bigger particles. The fractions of particles that are already large have been shown not to undergo sintering. The heat evolved from the exothermic sintering process causes the local nucleation of the rutile phase. Finally, as the conversion to rutile is also an exothermic process, this results in the transformation of the whole particle to rutile.

4.2.2 The effect of impurities on the anatase-rutile phase transformation

In many research works, researchers have observed that the inclusion of a certain amount of impurities into TiO_2 can drastically alter the physical properties of the film. It has been shown that silicon and phosphorus inhibit the transformation from anatase to rutile, with 100% anatase phase being retained at temperatures as high as 870°C for up to 3 h for thin films 80 K and 1500 K for bulk samples [29]. The retardation of the anatase-rutile transformation can be achieved with impurities [29]. Most researchers agree that oxygen vacancies are responsible for the overall transformation mechanism [30]. Thus, the oxides and fluorides that assist the transformation can substitute for Ti^{+4} in the anatase lattice, resulting in the creation of oxygen vacancies. On the other hand, the inhibiting effect of other impurities involves the reduction of oxygen vacancies due its substitution into the anatase lattice.

Titanium alkoxides are common TiO_2 precursors, with the most frequently used being titanium tetra isopropoxide (TTIP) (also called tetra isopropyl titanate). The residue of the organic binders results in carbon contamination of typically a few atomic weight percentage (at. wt.%), but as high as 13 at. wt.% being observed [31]. It is likely that carbon incorporation could be higher at low growth temperatures, as when higher temperatures were used the carbonate species decomposed, resulting in the removal of hydrocarbon fragments [32]. Titanium tetrachloride ($TiCl_4$) is another common TiO_2 precursor, and this results in chlorine contamination of the deposited film.

5. Importance and applications of TiO_2 thin films

Due to its interesting intrinsic properties, TiO_2 thin films have great importance and significance for a large variety of industrial applications. Titanium oxide, which belongs to the metal transition oxide family, was the most studied during the last two decades and demanded material in many fields of applications such as transparent electrodes, gas sensors, solar cells (PV), photocatalytic process, etc. To improve the performance of this oxide, doping TiO_2 with suitable metal ion dopants offers an effective method to adjust some of its physical properties. Generally, the doping of semiconductors with appropriate metallic element (Al, Nb, Sn, Ge, Fe, Ni and Cr) is one of the most effective ways in research for developing sensitivity applications such as photovoltaic solar cells, photocatalysis and pollution sensors. However, the interaction between the doping metals and the semiconductor is complicated because the interaction relates to the carrier concentration, defect level and surface states of the semiconductor, electronic, optical properties, and so on.

6. Synthesis of titania

The sol-gel technique is a suitable method for deriving nano-TiO_2 having unique metastable structure at low temperatures and excellent chemical homogeneity [33]. The novelty of the work is to synthesize nanosized (<25 nm) TiO_2 particles by optimized preparatory parameters associated with sol-gel method and characterize for their structural and optical behavior useful to photovoltaic application.

The synthesis process includes that the precursor TTIP, 3.5 mL, was slowly added to the mixture of isopropanol (25 mL), concentric hydrochloric acid (0.1 mL, catalyst) and distilled water (0.2 mL) with constant stirring for 30 min. The mixture undergoes hydrolysis reaction resulting in transparent pale yellowish solution. Then, the solution was allowed 24 h for gelation period. The gel was dried at 373 K for an hour and finely ground with mortar. Finally, the TiO_2 powder was calcinated at 673 K for an hour to obtain a nanosized particle with desired phase [34]. The different steps are involved in synthesis of nano-TiO_2 powder by sol-gel method and it was represented by the chemical reaction [35] in the system is

$$\begin{aligned} Ti^{4+}\left[-O-CH(CH)_3\right]_4 H_2O \overset{Isopropanal}{\underset{H^{\cdot}}{\rightarrow}} Ti\left(OH\right)_4 + 4[(CH_3)_2CH-OH] \\ \downarrow \Delta(450^{o}C) \\ TiO_2 + 2H_2O \end{aligned} \tag{1}$$

Characterization of TiO_2 powder was carried out by using X-ray diffraction analysis (**Figure 5**) with XPERT-PRO X-ray diffractometer in the range of 2θ values from 20° to 80° (λ = 0.1540 nm). The 7.7-nm-sized particles were determined from Debye-Scherrer's formula. The surface morphology of TiO_2 pellet (**Figure 6**) obtained using the VEGA3 TESCAN scanning electron microscope confirms nanosize spherical-shaped particles uniformly distributed without any aggregation and atomic force microscope AFM XE-100 topography images (**Figure 7**) exhibit the distribution of uniform spherical-shaped particles. The chemical compositions such as Ti (26.36 at. wt.%), O (68.5 at. wt.%) and C (5.07 at. wt.%) in the prepared TiO_2 powder were confirmed by the energy dispersive X-ray (EDX) spectroscopy (**Figure 8**). The absorption and transmission spectra were obtained using Perkin Elmer Lambda 35 UV/Vis Spectrophotometer and its transmittance is about 60% (**Figure 9**). The Fourier transform infrared (FTIR) spectra (**Figure 10**) were taken using SPECTRUM-RX 2 and represents the characteristic peaks in the range of wave numbers 4000–400 cm^{-1}.

The TiO_2 nanoparticles of size 7.7 nm have been prepared by optimized sol-gel technique. The XRD analysis reveals that the TiO_2 powder was highly crystalline (anatase phase) and nanostructured with tetragonal system. The SEM images exhibit the nanosized TiO_2 particles with less densification nature. The AFM study confirms the uniform distribution of spherical-shaped particles. The optical band gap of the TiO_2 is found to be 3.45 eV making it suitable for solar cell applications.

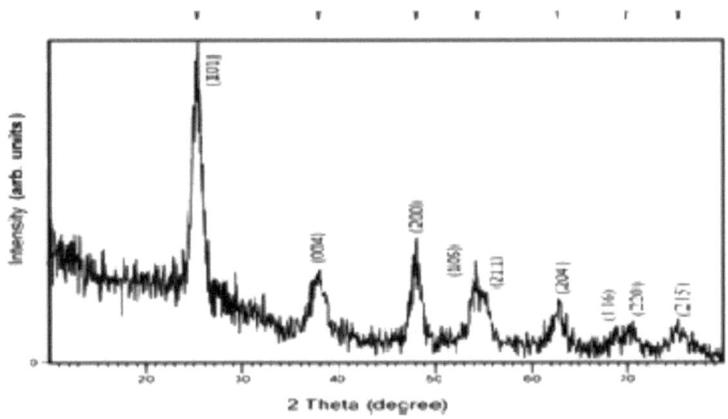

Figure 5.
X-ray diffraction pattern of TiO_2 powder calcined at 450°C.

Figure 6.
SEM micrographs of TiO_2 powder (a) before (b) and (c) calcined at 450°C with different magnifications.

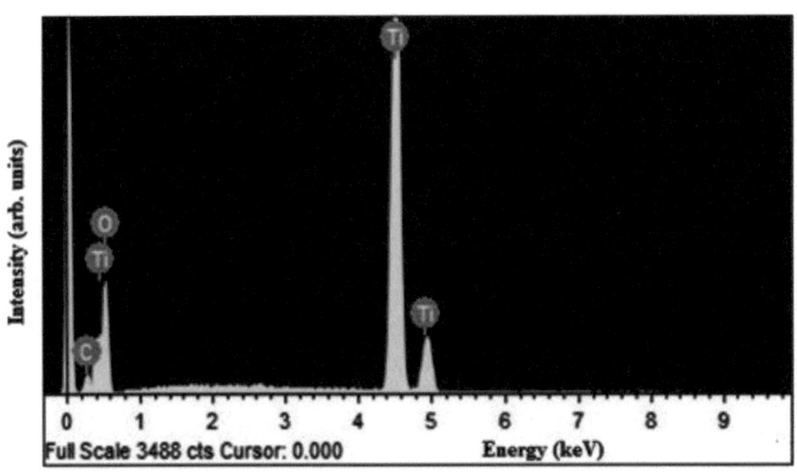

Figure 7.
EDAX spectra of TiO_2 powder calcined at 450°C.

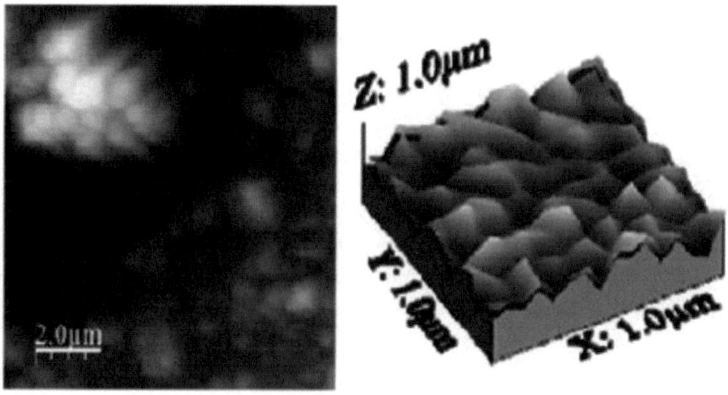

Figure 8.
AFM topographic images of TiO_2 powder calcined at 450°C.

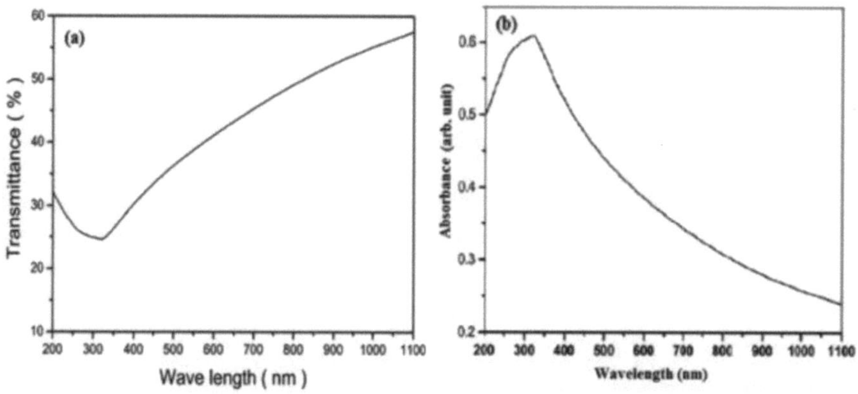

Figure 9.
UV (a) transmittance and (b) absorption spectra of TiO_2 powder calcined at 450°C.

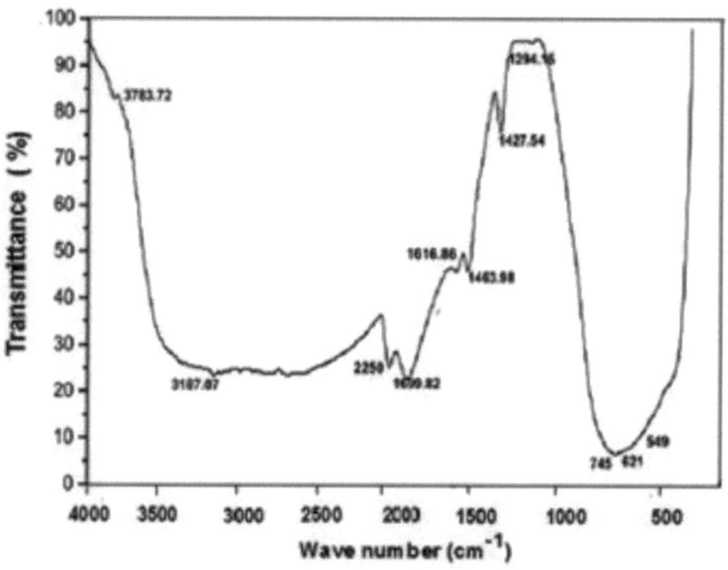

Figure 10.
The Fourier transform infrared (FTIR) spectra of TiO_2 powder calcined at 450°C.

7. Preparation of sol-gel routed nano-TiO_2 thin films and their effect of molarity

Nanocrystalline titanium dioxide (TiO_2) films are extensively studied because of their interesting chemical, electrical and optical properties. TiO_2 is one of the most important transition metal oxide semiconductors with wide band gap. A wide variety of techniques have been used to prepare titania films. Among these, the sol-gel routed spin coating technique has emerged as one of the most promising methods as it produces films by simple synthetic route with good homogeneity, low cost, excellent compositional control and feasibility of producing thin films on large complex shapes with low crystallization temperature.

This work is keeping the optimization of the processing parameters such as pH value (~8), amount of catalyst (HCl), spin speed (3000 rpm) and calcination temperature (450°C) constant to prepare nano-TiO_2 thin films with molar concentrations 0.05 M, 0.1 M, 0.15 M and 0.2 M by sol-gel routed spin coated technique. And also the study of the effect of molarity on structural, optical and electrical behaviors is useful to photovoltaic applications.

The titanium tetra isopropoxide (TTIP) was used as a precursor, hydrochloric acid as a chelating agent, isopropanol and deionized water as a solvent. Triton X-100 was used as a stabilizer to avoid precipitation in solution and at the same time used to increase the conductivity of films. TTIP (3.5 ml) was slowly added to the mixture of isopropanol (25 ml), concentric Hydrochloric acid (0.1 ml, catalyst) and distilled water (0.2 ml) with constant stirring for 30 min. Introduction of isopropanol prior to TTIP induces immediate precipitation due to highly reactive alkoxide, therefore Triton X-100 was added as a stabilizing agent for the hydrolysis reaction. The resultant alkoxide solution was kept at room temperature for hydrolysis reaction for 2 hours, resulting in a transparent pale yellowish TiO_2 sol. The hydrolysis and the poly condensation of titanium alkoxides proceeds according to the mechanism Eq. (1)). TiO_2 sol was deposited on to a glass substrate by a spin coating unit with spin rate at 3000 rpm for 60s

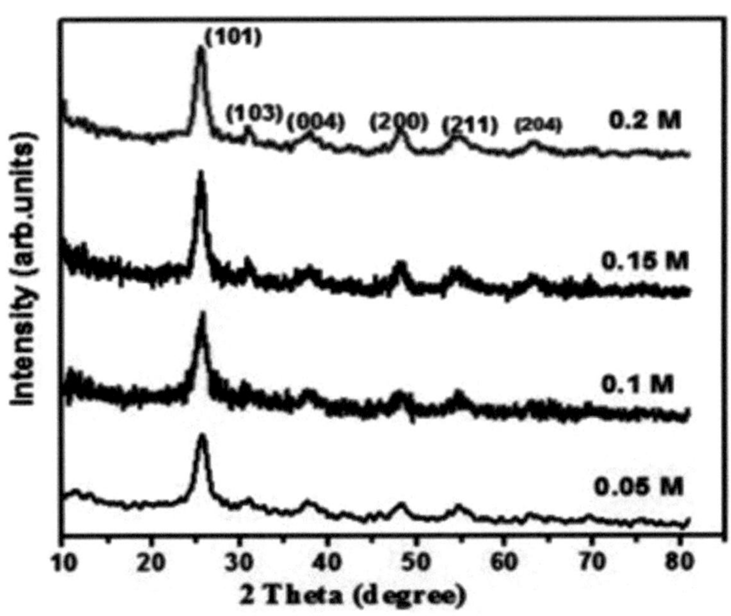

Figure 11.
X-ray diffraction pattern of TiO_2 thin films at different molar concentrations calcined at 450°C.

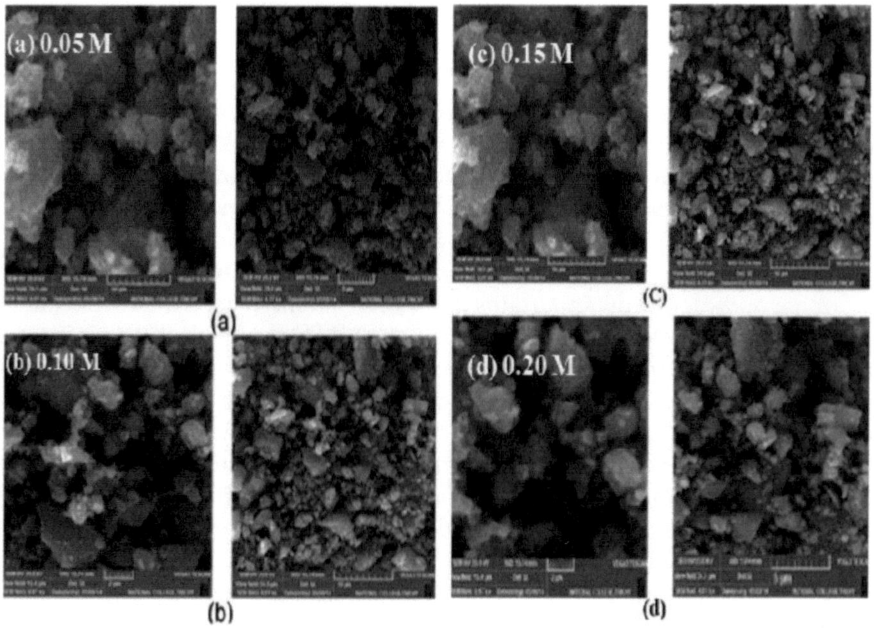

Figure 12.
SEM micrographs of doped and undoped TiO_2 thin films at different molar concentrations calcined at 450°C.

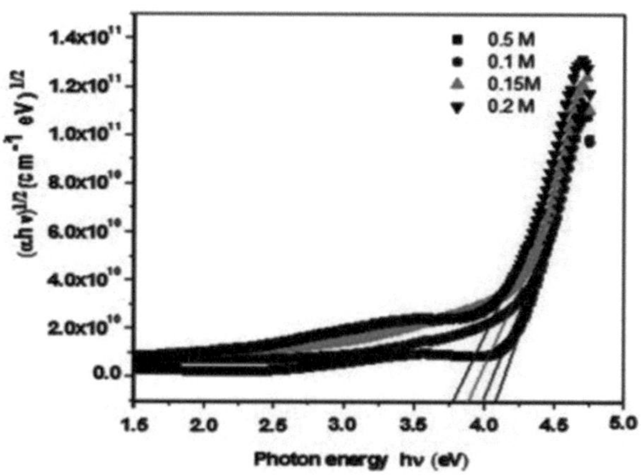

Figure 13.
Optical band gap energy of TiO_2 thin films at different molar concentrations calcined at 450°C.

in air and dried on a hot plate at 100°C for 60 seconds. The prepared samples were calcinated at 450°C for 1 hour [36].

The nanostructured titanium dioxide (TiO_2) thin films were prepared using the sol–gel routed spin coating technique. The X-ray diffraction pattern of TiO_2 thin films (**Figure 11**) exhibits that TiO_2 particles are crystallized as anatase phase and nanostructured with the tetragonal system. The SEM images (**Figure 12**) exhibit that the particles are spherical in nature. TiO_2 thin film prepared at 0.2 M concentration has a smooth surface. The roughness of the TiO_2 thin film increases with the increase of molarity. The optical transmittance is found to depend on the molarity and the higher value of molarity leads to lower optical band gap energy (**Figure 13**). Hence, the nano-TiO_2 thin films with higher molar concentration will be useful for photovoltaic applications due to their structural, optical and electrical behaviors.

8. Conclusions

Titanium dioxide (TiO_2) is one among the wide band gap metal oxide semiconducting materials used in thin films that have received much attraction due to its sensing properties, dielectric properties, antireflective coatings, good physical and chemical stability, high refractive index, low absorption, low cost, non-toxicity, high electron mobility, longer electron life time, low recombination losses and ease in preparation.

Titanium dioxide (TiO_2), also known as titanium (IV) oxide or titania, is the naturally occurring oxide of titanium. Nanoscaled titanium dioxide (TiO_2) in thin layer or nanopowder forms appertains to the most extensively studied semiconductors. This metal oxide is a promising semiconductor frequently used due to its non-toxicity, chemical stability, photocatalytic activity and low cost [37, 38]. Especially, thin films as the nanostructured electrode materials have become very important in the fields of photovoltaics, energy storage, sensing, photo-electro-catalysis, etc.

TiO_2 in crystallographic form of anatase has become an interesting candidate as an n-type photoanode due to its band gap (Eg = 3.2 eV), which is higher than that for the rutile phase (Eg = 3.0 eV) and has excellent efficiency to generate the electron-hole pairs [39, 40]. The preparation of the nanostructured electrode materials with highly uniform nanoparticles has been investigated by many groups [41]. The most commonly used method is the sol-gel technique utilizing the molecular templates. The main advantage of this purely chemical method lies in a possibility of layer preparation under laboratory conditions as well as the possibility to tailor TiO_2 layer properties by varying preparation conditions. Nanoparticles with controlled chemical composition, size distribution, uniformity and dispersion can be readily synthesized using reverse micelles [42].

The TiO_2 thin films have been introduced as electron transport layers (ETL) because of their large band gap (3.7 eV) and well-matched energy levels (valence band of ~ 8.1 eV and conduction band of ~ 4.4 eV). The requirements for ETL involve high electron mobility and transparency in the visible region to allow transmission of light into the active layer. These requirements limit the number of materials that have these characteristics, among which is the well-known and widely used titanium oxide [43].

Thus, in the present work, I have attempted to synthesis nanosized TiO_2 particles with 7.7 nm, which is less than commercially available TiO_2 powder (25 nm), and modified preparatory condition followed to prepare nanocrystalline TiO_2 thin films with different molar concentrations. The results from the above study exhibits that the prepared nano TiO_2 films with different molar concentrations are to enhance the optical, structural and electrical behavior of film which is suitable for photovoltaic applications.

Conflict of interest

The authors declare no conflict of interest.

Author details

Lourduraj Stephen
Department of Physics, St. Joseph's College (Autonomous),
Tiruchirappalli, Tamil Nadu, India

*Address all correspondence to: lourduraj82@gmail

References

[1] Standard CPDS X-ray Diffraction Spectra Database JCPDS Card No. 21-1272 (Anatase TiO_2), JCPDS Card No. 72-0100 (Brookite TiO_2), JCPDS Card No. 75-1753 (Rutile TiO_2)

[2] Staun Olsen J, Gerwald L, Jiang JZ. On the rutile/α-PbO_2-type phase boundary of TiO_2. Journal of Physics and Chemistry of Solids. 1999;**60**:229

[3] Dewhurst JK, Lowther JE. High-pressure structural phases of titanium dioxide. Physical Review B. 1996;**54**:R3673

[4] Haines J, Léger JN. X-ray diffraction study of TiO_2 up to 49 GPa. Physica B. 1993;**192**:233

[5] Mosaddeq-ur-Rahman MD, Miki T, Krishna KM, Soga T, Igarashi K, Tanemura S, et al. Structural and optical characterization of $Pb_xTi_{1-x}O_2$ film prepared by sol-gel method. Materials Science and Engineering. 1996;**B41**:67

[6] Gotic M, Ivanda M, Sekulic A, Music S, Popovic S, Turkovic A, et al. Microstructure of nanosized TiO_2 obtained by sol-gel synthesis. Materials Letters. 1996;**28**:225

[7] Aarik J, Aidla A, Sammelselg V, Siimon H, Uustare T. Control of thin film structure by reactant pressure in atomic layer deposition of TiO_2. Journal of Crystal Growth. 1996;**0169**:496

[8] D'Ans J, Ten Bruggengate P, Eucken A, Joos G, Roth WA. Landolt-Börnstein 6 II/8. Berlin: Springer-Verlag; 1965. pp. 2-145

[9] Eucken A, Buchner A. The Dielectric. Chemie (Leipzig). 1935;**27**(5-6):321-349

[10] Roberts S. Dielectric constants and polarizabilities of ions in simple crystals and barium titanate. Physical Review. 1949;**76**:1215

[11] D'Ans J, Eucken A, Joos G, Roth WA. Landolt-Börnstein 6 II/6. Berlin: Springer-Verlag; 1959. p. 483

[12] Ali I, Suhail M, Alothman ZA, Alwarthan A. Recent advances in syntheses, properties and applications of TiO_2 nanostructures. RSC Advances. 2018;**8**:30125-30147

[13] Grätzel M, Rotzinger FP. The influence of the crystal lattice structure on the conduction band energy of oxides of titanium(IV). Chemical Physics Letters. 1985;**118**:474

[14] Forro L, Chauvet O, Emin D, Zuppiroli L, Berger H, Lévy F. High mobility n-type charge carriers in large single crystals of anatase (TiO_2). Journal of Applied Physics. 1994;**75**:633

[15] Breckenridge RG, Hosler W. Electrical properties of titanium dioxide semiconductors. Physical Review. 1953;**91**:793

[16] Wahlbeck PG, Gilles PW. Reinvestigation of the phase diagram for the system titanium–oxygen. Journal of the American Ceramic Society. 1966;**49**:180

[17] Sze SM. Physics of Semiconductor Devices. New York: Wiley Interscience; 1981

[18] Kröger FA, Vink HJ. Solid State Physics. New York: Academic Press; 1956. p. 307

[19] Feuersanger AE. Titanium-dioxide dielectric films prepared by vapor reaction. Proceedings of the IEEE. 1964;**52**(12):1463-1465

[20] Fitzgibbons ET, Sladek KJ, Hartwig WH. TiO_2 film properties as a function of processing temperature. Journal of the Electrochemical Society. 1972;**119**:735-739

[21] Debnath R, Chaudhuri J. Inhibiting effect of $AlPO_4$ and SiO_2 on the anatase → rutile transformation reaction: An x-ray and laser Raman study. Journal of Materials Research. 1992;7(12):3348-3351

[22] Gennari FC, Pasquevich DM. Enhancing effect of iron chlorides on the anatase-rutile transition in titanium dioxide. Journal of the American Ceramic Society. 1999;**82**(7): 1915-1921

[23] Hass G. Preparation, properties and optical applications of thin films of titanium dioxide. Vacuum. 1952;**II**(4):331-345

[24] Takahashi Y, Tsuda K, Sugiyama K, Minoura H, Makino D, Tsuiki M. Journal of the Chemical Society: Faraday Transactions 1. 1981;**77**:1051-1057

[25] Guenther KH. Recent progress in optical coating technology: low voltage ion plating deposition. SPIE. 1990;**1270**:211-221

[26] Pulker HK, Paesold G, Ritter E. Refractive indices of TiO_2 films produced by reactive evaporation of various titanium–oxygen phases. Applied Optics. 1976;**15**(12):2986-2991

[27] Zhang S, Zhu YF, Brodie DE. Photoconducting TiO_2 prepared by spray pyrolysis using $TiCl_4$. Thin Solid Films. 1992;**213**:265-270

[28] Campbell SA, Kim H-S, Gilmer DC, He B, Ma T, Gladfelter WL. Titanium dioxide (TiO_2)-based gate insulators. IBM Journal of Research and Development. 1999;**43**(3):383-392

[29] Akhtar MK, Pratsinis SE, Mastrangelo SVR. Effect of dopants in vapor. Materials Research Symposium Proceedings. 1992;**271**:951-956

[30] Shannon RD, Pask JA. Kinetics of the anatase-rutile transformation. Journal of the American Ceramic Society. 1965;**48**(8):391-398

[31] Murozono M, Kitamura S, Ohmura T, Kusao K, Umeo Y. Titanium dioxide antireflective coating for silicon solar cells by spinning technique. Japanese Journal of Applied Physics. 1982;**21**(Supplement 21-2):137-141

[32] Babelon P, Dequiedt AS, M'ostefa-Sba H, Bourgeois S, Sibillot P, Sacilotti M. SEM and XPS studies of titanium dioxide thin films grown by MOCVD. Thin Solid Films. 1998;**322**:63-67

[33] Zhang J, Xiao X, Nan J. Hydrothermal-hydrolysis synthesis and photocatalytic properties of nano-TiO_2 with an adjustable crystalline structure. Journal of Hazardous Materials. 2010;**176**:617-622

[34] Chenga P, Denga C, Gub M, Dai AX. Effect of urea on the photoactivity of titania powder prepared by sol–gel method. Materials Chemistry and Physics. 2008;**107**:77-81

[35] Livage J, Henry M, Sanchez C. Sol-gel chemistry of transition metal oxides. Solid State Chemistry. 1998;**18-4**:259-341

[36] Lourduraj S, Williams RV. Structural and optical characterization of synthesized TiO_2 nanopowder using sol–gel technique. International Journal of Nanoscience. 2016;**15**:1650002

[37] Malato S, Fernandez-Ibanez P, Maldonado MI, Blanco J, Gernjak W. Decontamination and disinfection of water by solar photocatalysis recent overview and trends. Catalysis Today. 2009;**147**:1-2

[38] Butterfield IM, Christensen PA, Hamnett A, Shaw KE, Walker GM. Applied studies on immobilized titanium dioxide films as catalysts for the photoelectrochemical detoxification of water. Journal of Applied Electrochemistry. 1997;**27**(4):385-395

[39] Hashimoto K, Irie H, Hujishima A. TiO_2 photocatalysis: A historical overview and future prospects. Japanese Journal of Applied Physics. 2005;**44**:12-8269

[40] Gracia F, Holgado J, González-Elipe A. Langmuir. Photoefficiency and optical, microstructural, and structural properties of TiO_2 thin films used as photoanodes. Langmuir (ACS). 2004;20:5-1688

[41] Ying L, Hon LS, White T, Withers R, Hai LB. Controlled nanophase development in photocatalytic titania. Materials Transactions. 2003;**44**:7-1328

[42] Li M, Schnablegger H, Mann S. Coupled synthesis and self-assembly of nanoparticles to give structures with controlled organization. Nature. 1999;**402**(6760):393

[43] Yeo KS, Nakaob S, Hirosea Y, Hasegawaa T, Matsuo Y. Application of sputter-deposited amorphous and anatase TiO_2 as electron collecting layers in inverted organic photovoltaics. Organic Electronics. 2013;**14**(7):1715-1719

Chapter 4

An Overview of Anodic Oxides Derived Advanced Nanocomposites Substrate for Surface Enhance Raman Spectroscopy

Marta Michalska-Domańska

Abstract

Surface-enhanced Raman spectroscopy (SERS) is an analytical technique, which allows to identify traces of chemical or biological substances in many field, like pharmaceutical and food industries, homeland security, nanosensors, or environmental protection. The analytes are identified based on their vibrational spectra, unique for a given compound. The advantage of SERS is effective qualitative analysis of trace amounts of analyte, but the disadvantages are stability of substrate and repeatability of measurements. The challenge is to improve SERS substrates to minimize these drawbacks. Nowadays high-precision electron beam lithography or focused ion beam is used in SERS substrate fabrication, which is impractical for large-scale production. In recent years, researchers' attention has focused on porous anodic oxides, with inexpensive and scalable production method, as potential materials for SERS substrates. This chapter will discuss the progress of anodic oxides used as a SERS substrate, and the brief description of conventional SERS substrate fabrication methods will be presented.

Keywords: nanocomposites, anodization, anodic oxide, titanium anodic oxide, aluminum anodic oxide, porous, plasmon, hot spots, surface-enhanced Raman spectroscopy, SERS substrate

1. Introduction

Raman spectroscopy is a technique based on measuring Raman scattering, i.e., inelastically scattered photons. It belongs to the techniques of testing the oscillatory spectra of materials. Thanks to its application, it is possible to obtain information about the type and structure of various chemical and biological compounds, including explosives, drugs, molecules of poisonous compounds, as well as bacteria, viruses, and even cancer cells. This technique has a lot of advantages, but the biggest disadvantage is the low sensitivity of the method, which can be solved by the use of surface-enhanced Raman spectroscopy (SERS). It increases the intensity of the Raman signal up to 10^{14} times compared to the signal received in a classic Raman experiment [1]. The first observations of the surface-amplified Raman signal for pyridine were made on the rough silver surface as early as 1974 [2]; however, the authors did not recognize the observed phenomenon as an extraordinary amplification or phenomenon. Since the discovery of SERS in 1977 [3], interest in this research technique has grown exponentially.

SERS is a well-known technique for detecting trace amounts (or even individual molecules) of chemical and organic substances in real time, using a specific structural and vibration "fingerprint" of the analyte being tested. By using this technique, it is possible not only to efficiently qualitatively analyze the particles adsorbed on the surface of the substrate but also to obtain valuable information on the structure of the detected compounds. SERS can be used, among others, in medicine, pharmaceutical and food industry, internal security, the detection of terrorist threats, nanosensors, or environmental protection [1, 4].

SERS is a noninvasive technique; it can be used for solid, liquid, and gas samples and does not require sample preparation before measurement; and it does not require vacuum. In biological analysis, it is possible to work in vitro and in situ. Analyses can be carried out in a wide range of measurement conditions (temperature, pressure) and are short-lived (lasts for seconds); a small amount of the analyzed substance (up to one molecule) is sufficient. It is possible to detect many components of a sample at the same time because the spectra obtained are characterized by high resolution.

A few years ago, SERS measuring equipment (based on confocal microscope, monochromator, and detector—cooled silicon matrix) was expensive and has significant dimensions. Nowadays, there are more and more inexpensive, portable technical solutions with increasingly better parameters (although their spectral resolution is still relatively low). Despite the advantages, one should take into account the limitations of this research methodology: problems with quantitative analysis, problems with the determination of fluorescent materials, and the inability to determine metals and alloys, as well as the stability and repeatability of measurements. The latter one problem can be solved by improving the SERS substrates used.

SERS substrate can be classified by structure of substrate surface, fabrication method, or application. According to the used structure, the SERS substrates can be divided based on:

a. Metal nanoparticles (e.g., colloids)

b. Metal thin films

c. Nanowires

d. Composites

Generally, all fabrication methods of the SERS substrate production can be divided into four main categories [5]:

a. Chemical synthesis (like synthesis of colloids by microwave heating method) (1)

b. Deterministic pattering (including all lithographic techniques, like electron beam lithography and focused ion beam fabrication) (2)

c. Template-assisted methods (3)

d. Other methods not mentioned above, which can be classified as nonconventional (4)

The two first categories (1, 2) can be classified as a "bottom-up" approach, while the third (3) is typical "top-down" approach [5]. The fourth category (4) collects all new technologies, which present new proposal and are based on new ideas, not used traditionally. For example, when anodic aluminum oxide (AAO) is used as a template for nanofabrication of metallic nanostructures, one can classify that method to the third category (3), while AAO decorated by metallic nanoparticles and used as an active SERS substrate will be included in nonconventional method (4). In summary, the classification of the fourth category depends on what experimental actions are included there, so the decision whether it is "bottom-up" or "top-down" approach should be taken after the analysis of the method.

SERS substrate can be divided according to application:

a. Military utilization and internal security, e.g., detection and identification of illicit or dangerous materials such as explosives

b. Quality control in the food and pharmaceutical industry

c. Medicine, e.g., detection of cancer cells or defects in DNA

d. Biology

e. Environmental science

f. Single molecule detection

To describe SERS substrate, the following parameters have to be taken under consideration: enhancement factor of signal for a given analyte, repeatability of measurements, shelf life/durability, required excitation wavelength and laser power range, temperature, and humidity sensitivity. To define SERS substrate efficiencies, the specific application, where the substrate is planned to be used, should be included. For example, when we plan to use SERS in blue or green-orange region (cover visible and NIR light), we can take an Au or Ag nanocolloids, but for UV region we need to use Al nanoparticles [6]. Generally, for detection of microorganisms and living cells, the colloidal nanoparticles are typically preferred as SERS substrates. Also, for detection of specific sequence of DNA or RNA and Ag or Au nanoparticles, Raman-active dyes can be applied, while in detecting pancreatic cancer biomarkers, the SERS in body fluids is required [7, 8]. Detection of bacteria *E. coli* can be made on SERS substrate with gold nanoparticles [9]. For detection of explosive materials, traditionally silver- or gold-based nanostructure is typically used [10–12], but Cu-based nanoobjects fabricated by laser ablation were also tested [13]. Another important factor that influenced the Raman enhancement is the environment, in which the detected molecules are embedded, such as body fluid, ethanol water, or other solvent.

Enhancing Raman scattering over ordinary ones, spectrum registration conditions are on the order of 10^6 or greater. It is assumed that the observed signal enhancement is the resultant of two mechanisms: chemical strengthening and electromagnetic field amplification.

Generally, the Raman enhancement depends on electromagnetic and chemical amplification in the presence of metallic nanostructure (Ag, Au, Cu, Al, etc.). Chemical strengthening arises as a result of the overlapping of valence orbitals of the adsorbed molecule and the metal conductivity bands. Then a charge transfer from the adsorbate analyze to the metal (or vice versa) becomes possible. The strength of the SERS enhancement is strongly dependent on the shape and size as well as dielectric constant of the metal (plasmonic) nanoparticles on SERS substrate [14, 15], because this affected the ratio of absorption and scattering events. When the particles are too large, the excitation of nonradiative multipoles can appear, and a decrease in the overall efficiency of the enhancement may occur. On the other hand, too small particles lose their electrical conductance and did not enhance the field. When the particle size move toward few atoms, we do not have plasmon anymore at the surface, because we do not have large collection of electrons oscillated together. Each experiment has an ideal particle size and ideal surface thickness to achieve best performance [16]. The strength of chemical enhancement is much lower than electromagnetic ones, but it is very important in determination of spectral pattern of SERS spectra, i.e., Raman shift or intensity ratio [17]. When the interaction between molecule and the metal produces a metal-molecule charge transfer (CT) state and the Raman scattering is excited with a laser source in resonance with that state, some Raman modes can be strongly enhanced. In summary, the critical issue for the improvement of SERS sensitivity and reproducibility is the rational design of a stable and uniform SERS-active substrate, because the Raman enhancement is susceptible to the composition, size, and morphology of the substrate [18–20]. Higher SERS substrate surface area favors obtaining enhanced higher signal, and for this reason the work is underway on obtaining SERS substrates constructed of nanostructures coated with metallic plasmonic nanoobjects.

The high local electromagnetic field near plasmonic nanostructures, when the tested molecules are in their immediate vicinity, provides enhanced high Raman signal. The degree of electric field enhancement depends on the composition, morphology, and geometric parameters of surface nanostructures, as well as the amount of the so-called hot spots (regions of intense local field enhancement believed to be caused by local surface plasmon resonances (LSPR) [21]), and decreases sharply as the distance between the ground and the analyte under test increases above a dozen nm. There are a lot of publications described in detail in the LSPR phenomenon [22–28], which justified not taking up this issue in this chapter.

Plasmon hot spots spread over the entire surface of the SERS substrate to ensure high sensitivity and repeatability of spectroscopic measurements; however, several conditions must be filled: (1) homogeneous distribution of plasmon nanostructures on a large surface with a strictly controlled distance, which effectively produces hotspot points; (2) it is necessary to efficiently and homogeneously deposit the studied molecules on the entire SERS substrate [29].

The main features of both the abovementioned mechanisms of SERS enhancement are different and will be given below. Chemical reinforcement can be characterized as follow:

- The enhancement of Raman scattered light intensity in relation to the usual Raman spectrum registration conditions is of the order of 10^2.

- Enhancement occurs only for particles that interact directly with the metal—as a consequence the range is limited to the monolayer adsorbate (high surface specificity).

- Chemical enhancing does not depend on the optical properties of the metal but on the nature of the metal-adsorbate interactions; for that reason they are also observed for other metals than traditionally used in SERS, like Ag, Au, or Cu.

- The amount of chemical enhancing depends on the location of the Fermi level of the metal and changes depending on the applied electrode potential.

Electromagnetic field enhancement can be characterized as follow:

- The field enhancement effect depends mainly on the optical properties of metal and is the strongest for Ag, Au, and Cu
- Field amplification increases the scattering intensity of Raman order of 10^4 in relation to normal condition registration of Raman spectrum.
- This enhancement is felt up to several dozen nm from the surface of metal (range covers several atomic layers).
- The effect strongly depends on the size, shape, and roughness, the surface morphology of the metal, or the size of the aggregates of metal atoms.

To be able to take full advantage of the possibilities of the SERS technique, and even expand its application, it is necessary to develop effective SERS substrates that will provide Raman dissipation signal enhanced up to 10^6 and larger. These SERS substrates have to be time stable and mechanically durable and ensure repeatable analysis results. To realize this aim, the SERS substrates based on nanostructures, including metallic nanostructures, have to be developed. The main trends in the matter of searching suitable nanostructures are as follow:

- Colloids
- Layers of metals
- Layers covered by nanoobjects
- Nanostructures covered by metallic nanoobjects/plasmonic nanoparticles
- Mix of form mentioned above

Brief descriptions of conventional fabrication methods and new directions to production of nanostructures for application in SERS are presented as follows:

In recent years ideas and attempts of using the anodic aluminum oxide and anodic titanium oxide as SERS substrate were made. The ideas can be divided into the following groups:

1. Using AAO as a matrix for the nanofabrication of nanostructures (like nanodots or nanowires) which can act as SERS substrates
2. AAO coated by metals (mainly by silver, AAO/Ag) in many forms directly used as the composite SERS substrate
3. ATO coated by metals (Ag or Au) which can act as recyclable SERS-active substrate

More detailed information about research of anodic oxides using as SERS substrate will be given in Part 4 of this chapter.

In this chapter, the main emphasis will be on a new type of SERS substrates based on anodic metal oxides, which is the perspective and future direction of SERS substrates. Generally, AAO and ATO consist of arrays of regularly arranged nano-tubes/nanopores, which can be filled or covered by other materials. Moreover, ATO after annealing can act as a semiconductor and be used, for example, as a photocatalyst for removal of residual particles left on the substrate after SERS measurements. Due to the listed above advantages, the anodic oxides in SERS applications become more and more popular but up to now only in the research area. It is worth to notice that aluminum and/or titanium anodic oxides as a SERS substrate are no commercially produced now, so as a consequence its fabrication methods can be classified as unconventional fabrication SERS substrate method. The possibilities of application of anodic aluminum oxide and anodic titanium oxide as SERS substrates will be described and discussed in details below.

2. Conventional SERS substrate fabrication methods

Conventional SERS substrate types are nanostructures covered by metallic nanoobjects (NS/PLA), including nanoparticles or colloids. The fabrication methods of that substrate are briefly summarized below.

The SERS substrates in a type of nanostructures covered by metallic nanoobjects (NS/PLA) are produced by the application of the physical vapor deposition (PVD) technique of thin discontinuous metallic layers on nanostructures [30], using lithographic techniques (photolithography, electron or ion beam lithography) [31], nanoprinting or lithography supported by a template (matrices of porous polymers or shadow lithography—nanosphere lithography (NSL)) [31]. However, the methods listed above are relatively expensive and time-consuming for areas larger than the range of mm^2 [32]. The limitation of using lithography method in SERS substrate production is connected with fabrication cost of nanostructures on large area [33] and resolution [34] of this technique. Below, in **Table 1**, the main disadvantages of that approach are presented.

The use of Au, Ag, or Cu metal colloids, with a diameter in the range of 10–80 nm, which are produced by chemical reduction, requiring the use of expensive chemical reagents, but the largest enhanced SERS (up to 10^{14}), is shown for highly aggregated colloids [31]. The biggest disadvantages of using colloids are storage (usually they should be used immediately after preparation), difficulties in use in portable spectrometers, high cost, problems with repeatability, and the limited range of materials that can be determined due to colloids' dimensions. The use of colloidal metallic particles or nanowires on a glass substrate ensures a relatively homogeneous size of nanoobjects; however, their placement on the substrate is random and unique.

Since few years, the interest on new kind of flexible SERS substrates grows linearly. The new interesting group of SERS substrates constitutes flexible and highly sensitive multiparticle complexes. They are mainly based on metal nanoparticles

Advantages	Disadvantages
• The possibility of controlling the metal nanoparticles' size and shape as well as the interparticle distance with great accuracy • Good reproducibility at a molecular resolution range	• Expensive equipment • Many steps are required to prepared one substrate • Quite long processes, so long manufacturing time for higher areas than nm^2 • High cost for areas larger than few μm^2 • Mass production is doubtedly

Table 1.
Main positive and negative features of using lithography methods in SERS substrate production.

combined with different kinds of fibers, thanks to which they possess new properties. To prepare that kind of SERS substrate, many approaches are taken under consideration, including the use of polymer nanofabric [35, 36], cellulose [37, 38], or flexible piezoelectric-modulated layer [39] decorated by silver or gold nanoparticles or nanowires. Few of these substrates can find its application in food industry, especially for pesticide detection [37, 38, 40], and others in explosive materials detection [41]. Despite many unquestionable advantages, such as high sensitivity, that kind of substrates can be temperature and humidity sensitive as well as still has short lifetime. In this chapter other issues related to this type of SERS substrate will not be addressed.

3. Commercially available SERS substrate

There are several commercially available SERS substrates. They are manufactured based on silicon etching and metal deposition techniques (Klarite substrates (discontinued), Renishaw), colloid distribution on cellulose fiber substrate (Ocean Optics substrate), or ion etching (Silmeco substrates). For more information on these solutions, see **Table 2**. All data were taken from the companies' websites.

Company name/SERS substrate name	Fabrication technology	Features	Price/active area
Renishaw-Klarite substrate (production terminated)	Silicon etching and coating of 300–500 nm Au	• High repeatability of microstructures • SERS enhancement was not high	Around 75 EUR/ piece (4 × 4 mm)
Diagnostic anSERS Inc., USA	Colloidal Ag nanoparticles deposited on cellulose fibers, printed	• Durable substrates • Restrictions on use in the excitation wavelength and laser power range • Single-use medium • Gain up to 10^6 depending on the analyte • Temperature and humidity sensitivity	175$/ 25pieces (0.25 inch)
Ocean Optics Inc., USA	Colloidal Ag nanoparticles deposited on cellulose fibers—printing	• Single-use • Durability 1–1.5 months • Temperature and humidity sensitive • Restrictions in the excitation wavelength and laser power range	64$/piece (cellulose/Au) 62$/piece (cellulose/Ag) (5.5 mm)
Ocean Optics Inc., USA	Metallic Au or Ag nanosponges	• Reusable • Durability up to 6 months • Restrictions in the excitation wavelength and laser power range	80$/piece, Au or Ag nanosponge (4 × 4 mm)
Silmeco ApS, Denmark	Si nanostructures coated by Ag or Au, ion digestion and vapor deposition, solution deposition or incubation	• Laser power restrictions • Relatively high cost of substrate production	350$/5pieces (4 × 4 mm)

Company name/SERS substrate name	Fabrication technology	Features	Price/active area
Horiba Ltd., Japan	Nanobludgeons, vapor deposition, dynamic evaporation at an angle	• Measurement recommendation on a spherical area 4 mm in diameter at the center of the substrate	No information about the price (4 × 3 mm or 5 × 7 mm)
Enhanced Spectroscopy, Inc., USA	Ag layers, physical vapor deposition	• Very low durability, up to 70 hours after opening the container • 10^5–10^7 enhancement, depending on the analyte type	No information about the price
Nanova Inc., USA	Au nanoparticles with bimodal diameter distribution (60 nm and 15 nm)—larger particles are probably colloids and smaller ones are probably vapor deposited	• Only a laser with a wavelength of 785 nm • Long shelf life • The surface is not homogeneous, which results in lower repeatability of analyses	25$/piece

Table 2.
Commercially available SERS substrates with company name, main features, and oriented price.

4. Anodic oxides used as a SERS substrate

Anodic oxides are the group of materials which can be described as layered material made on the top of valve metal (like Al, Ti, Zn, Sn, W, etc. [42–47]) or its alloy (like Ti6Al7Nb, FeAl, etc. [48–51]) by electrooxidation, called anodization. Nowadays it was discovered that it is also possible to anodize cupper foil which leads to production of anodic oxide layer which consists of nanobundle or nanorods [52–54]. Anodization is the only method for anodic oxide production. The main morphologies possible to obtain by anodization are:

- Nanopores
- Nanotubes
- Nanowires/nanofibers

The most popular anodic oxides are aluminum anodic oxide (AAO) and titanium anodic oxide (ATO).

AAO is a layer of nanocapillaries, perpendicular to the Al substrate, with a diameter in the range of nanometer scale (nanoporous morphology).

Two main applications of AAO are:

1. Used as a template for nanofabrication of AAO made in two-step anodization process, when the nanocapillaries are hexagonally arranged [55–57]

2. Used as a protective, anticorrosion layer on the top of aluminum alloys [58, 59]

ATO can be made by anodization in two morphologies: nanoporous (similar to AAO, but in worse nanopore arrangement) or nanotubular [60–62].

The morphological features of ATO can be relatively easily controlled by parameters of titanium anodization process. ATO is mainly used as a photocatalyst (after annealing) [63], but new applications of that material are still developed. Below, the way of using AAO and ATO as a SERS substrate will be discussed.

The nanofabrication of nanostructures using the anodic aluminum oxide (AAO) template is one of the methods leading to the production of uniformly distributed nanoobjects on the substrate. The AAO is system of hexagonally ordered nanocapillaries, which are vertically oriented to the substrate surface. AAO matrix is produced during an electrochemical oxidation of aluminum foil. It is possible to control geometric parameters of AAO (such as pore diameter, distance between them, matrix thickness, degree of nanopore ordering, etc.), implemented by appropriate selection of matrix production process parameters (e.g., type, concentration and temperature of electrolyte, applied voltage, anodization time) [64–71]. It provides wide possibilities of controlling the morphology and distribution of metallic nanostructures fabricated by AAO template-assisted method, which provoked the multiple use of the AAO matrix in research aimed at making SERS substrates. Below one can find examples of using AAO in preparation of SERS substrate.

Because of the advantages of using AAO in nanostructure fabrication (like low cost, facile, and recurrent), a lot of research groups study the possibility of utilizing these materials in SERS substrate production. An ultrathin AAO membrane (up to 100 nm thick) was used to produce nanodots, nanowires, or metallic nanowires with plasmonic properties, and samples prepared in this way were used as SERS substrates [72–79]. Successful attempts were also made to produce SERS flexible substrates using AAO matrices [20, 80, 81]. AAO was also used as a matrix for the production of composite SERS substrates [82, 83] or after coating with silver (AAO/Ag) directly as the SERS substrate [84–86].

Nguyen et al. [73] built SERS substrate based on hexagonally ordered Au nanodisks, which was deposited on quartz by ultrathin AAO used as a shadow mask. The tested molecule was methamidophos (MAP) at various concentrations ranging from 1 ppb to 10^3 ppm. It was found that SERS signal depends on the morphological features of nanodisks: diameter and distance between nanostructures. The authors observed clearly the tendency of shifting the plasmon resonance to the longer wavelength with increasing the size of Au nanodisks and with decreasing the distance between them.

Liu et al. also reported correlation between the size of Au nanocones and its SERS properties in rhodamine 6G (R6G) detection [74]. They prepared highly ordered Au nanocone arrays with a diameter in the range of 36–77 nm and the distance between the centers of nanocones about 100 nm. To prepare this substrate, the AAO mask was used and Au was evaporated on glass. It was experimentally proven that the spectral position of plasmon resonances was seen to slightly blueshift with increasing the nanocone size, which coincided well with the result of finite-difference time-domain (FDTD) simulation. Moreover, compared with the bulk sample, the estimated SERS enhancement factor in R6G detection was boosted from 8.00×10^6 to 1.79×10^7 as the diameter of Au nanocones increased from 36 to 77 nm [74].

Kim et al. [76] fabricated a highly ordered array of Au nanorod arrays consisted of agglomerated nanoparticles by porous anodic aluminum oxide (AAO) template-assisted electrochemical deposition. That prepared SERS substrate was tested for 1 mM 4-methylbenzenethiol (4-MBT) detection in three systems: Au nanorod arrays with agglomerated structures after removing AAO template, smooth structures after removing AAO template, and agglomerated structures before removing AAO template. It was shown that the Au nanorod arrays with agglomerated structures demonstrated a higher activity in SERS effect as abundant nanogaps are created uniformly by combination of hot spots caused by both agglomerated porous structures on each nanorod and inter-rod gaps. Interestingly, that SERS substrate tested before removing AAO template exhibits the lowest signal intensity.

Liu et al. described a universal method to fabricate large-area regularly ordered gold nanodots using an AAO mask [79]. The Au nanodots were patterned on different substrates such as silicon wafer, gold, and graphene substrates, and then the films were used as SERS-active cathodes in Li-O_2 batteries. The discharge products on the different electrodes (graphene and gold) were analyzed, and the results indicated that the cathode reactions on these two electrodes are distinct from each other. The SERS electrodes are so sensitive that even a small amount of product formed on the nonporous, planar electrode can be detected clearly. The authors ensured that SERS-active electrode provided a uniform enhancement of the Raman spectra over the entire substrate, which is more suitable for detecting the intermediates and by-products formed on the electrode.

Several research groups work on developing flexible SERS substrates based on polymer and metallic layers, which are fabricated with the use of AAO membranes as a template for giving the shape to polymer. Zhang et al. [20] tested flexible SERS substrate made by Au sputtering on the surface of the polymer nanostructure arrays, which was made by R2R imprinting UV-NIL process where AAO was used as the mold. The thickness of sputtering Au layer was optimized, and finally the substrate with a 30-nm Au coating was chosen as it showed the highest SERS enhancement for R6G detection. The largest enhancement factor for R6G at 1366 cm^{-1} was calculated at 1.21×10^7, and the SERS performance showed no obvious differences under different bending angles and different bending cycles. The tested substrate demonstrated excellent SERS performance and flexibility.

Zhao et al. [80] have transferred three-dimensional nanoparticle array to polymer film with the use of AAO mask. As a result, polymer-nanocone-based 3D Au nanoparticle array SERS substrate was obtained. The new class of SERS substrate presented displays a high SERS sensitivity to R6G in a concentration up to 10^{-12} M. The enhancement factor was calculated at 1.30×10^8. In the other paper, Zhao et al. have reported fabrication of polymer nanopillar array with Au nanoparticle inlays used as a flexible and efficient SERS substrate [81]. The substrate was prepared by nanoimprint lithography (NIL) method using an anodic aluminum oxide (AAO) template. It was found that substrate shows high SERS sensitivity to R6G identification with concentration up to10^{-12} M and the enhancement factor was calculated at 8.20×10^7. Moreover, both tested SERS substrates [80, 81] exhibit remarkable reproducibility, excellent flexibility (the SERS intensity acquired from the substrate almost remains constant after 200 bending times), and transparency in visual and infrared range, are lightweight and can be easily handled.

Chen et al. have used AAO with silver nanoparticles entrapped in the nanopores as a SERS substrate [86]. The Ag nanoparticles were seeded growth from aqueous $AgNO_3$ solution inside the AAO pore channels and treated at elevated temperature. It was shown that Ag nanoparticles entrapped in the pore channels exhibit significant structural stability and are less susceptible to oxidation than Ag NPs on the surface of a planar substrate at high temperatures. It was revealed that Ag NPs in AAO retain substantial SERS activity after heat treatment at 500°C for 5 days and even 600°C for 3 hours. The authors claim that geometric confinement of Ag NPs in the pore channel structure of AAO has clearly played a significant role in endowing them with the enhanced stability. It seems to be likely that the stability region may well be expanded to higher temperature and longer durations with reduced dimensions of the pore channels and decreased porosity of AAO. The SERS activity of tested samples with 10^{-6} M R6G was studied. It was shown that SERS spectra made on as-prepared and annealed samples are basically the same, which suggests that the coalescence of adjacent Ag NPs occurring at this stage has insignificant impact on the overall hot spots available for SERS enhancement.

Lim and Ng [87] compared the SERS enhancement level for 2-naphthalenothiol obtained on AAO nanofibers coated with a thin Ag layer and on a metallic thin Ag layer deposited on nanoparticles and showed that the SERS signal enhancement on a properly selected length of AAO nanofibers is almost three times higher than on a thin Ag layer.

Dan et al. used AAO as a template for Ag nanorod array fabrication and then removed the oxide and decorated nanorods by Au nanoparticles [88]. As-prepared substrate was used as a SERS substrate for R6G. The detection limit is as low as 10^{-16} M with excellent recyclability. According to the authors, the Raman enhancement occurs in the gap between the nanorods and the top of the nanorod bundle. The authors believe that this synthesis method can be used to prepare other metallic materials with outstanding structure, which can be used in many fields.

Zhong et al. [89] reported fabrication of Fe nanorod arrays by AAO template-assisted method and covered them by Au nanoparticles by sputtering approach to produced large-area urchin-like Au nanospheres for testing as SERS substrate for rhodamine 6G detection. The lower concentration of detected R6G with the use of as-prepared SERS substrate was 10^{-12} M. Moreover, this substrate exhibits excellent reusability after being cleaned. Based on the analysis of three-dimensional finite-difference time-domain (FDTD) simulations, the strong SERS activity is mainly due to the high density of hot spots induced by the gaps between the neighboring Au-NSs. According to the authors, the urchin-like arrays exhibit promising potential as new platforms to realize novel optoelectronic devices and have a great potential for more application in other fields.

Despite the AAO popularity, scientists also found interest within the ATO matrix as a production part of composite SERS substrate. Arrangement of ATO nanopores or nanotubes is not so regular like in AAO, but crystal form of TiO_2 shows photocatalytic properties, which can be of advantage in construction of reusable SERS substrates. Recently, there have been reports of nanocomposites composed of semiconductor and noble metal, e.g., ZnO/Ag [90] and TiO_2/Ag [91–94] or protected by the patent of Polish inventors GaN/Ag [95, 96], used as SERS substrate. Below possibility of composite SERS substrate build of anodic titanium oxide and metallic nanoparticles will be discussed.

Jakubowicz et al. [93] prepared ATO membranes with Ag nano-trees embedded in the pores. They found that the Ag nano-trees have a uniform branched symmetry independent of the applied substrate. In the cited paper, the SERS activity has not been studied; however other groups reported production of similar Ag nano-trees in AAO membrane and used this as a SERS substrate for R6G identification [84]. The SERS detection limit for R6G was estimated to 1.00×10^{-11} M. Moreover, unlike the case of conventional modification of hydrophobic functional groups, the Ag dendrite-integrated AAO membrane substrates have potentials in simple, rapid, direct, and sensitive detection of pollutants with low affinity for noble metal surface, such as fluoranthene. The authors ensure that Ag-dendrites/AAO membrane exhibited excellent SERS performance with high sensitivity, good reproducibility, and tenability [84].

Sun et al. [97] recognized Ag-coated ATO as a low-cost, uniform, and recyclable SERS-active substrate for 2-mercaptobenzoxazole (MBO) detection. The tested composite consisted of ATO made by two-step anodization of Ti foil with deposited Ag nanoparticles on the top of ATO by e-beam evaporation. The clear SERS signal was observed for very low concentration of MBO like 10^{-8} M. Utilizing photocatalytic properties of titanium dioxide, the SERS substrate after using was illuminated by UV light by 20 minutes and rinsed in distilled water to remove residual ions and molecules and dried at room temperature. SERS substrate cleaned in that way was reused for MBO detection with success.

Wen et al. described ATO coupling with Au nanostructures as a reliable SERS substrate with application in medicine, for chronic myeloid leukemia drug evaluation [98]. A critical biomarker of apoptosis, caspase-3, was detected by SERS analysis in real samples with reasonable recoveries. The authors assumed that SERS substrate developed by them has a lot of advantages (like cost-effectiveness, excellent reproducibility, and high sensitivity), which endows it with promising potential in apoptosis monitoring and anticancer drug development.

Ling et al. reported the use of Ag embedded in TiO_2 nanotube array as a recyclable SERS substrate for R6G molecule detection [99]. Concentration of R6G in SERS measurements was 10^{-6} M. It was found that the morphology of deposited Ag is strongly affected by the diameters of ATO nanotubes and the UV irradiation induced Ag aging process, especially the self-migration of silver along the tubular wall. The SERS activity is not proportional to the ATO diameters, and the enhancement factors on various roughened metallic surfaces strongly depend on the size, shape, distribution, and interaction of Ag nanoparticles with ATO.

In the other research, Roguska et al. [100] investigated Ag, Au, and Cu nanoclusters deposited on TiO_2 nanotubes/Ti as SERS substrate for pyridine detection. The influence of ATO nanotube diameters as well as the amount of deposited metal on SERS performance was studied. It was found that SERS activity of composite substrate was strongly dependent on the amount of deposited metal on ATO; for example, above 0.06 mg Ag/cm^2, the SERS intensity signal was higher than for referential substrate consisting of bulk activated Ag. Moreover, this high SERS activity of tested substrate, according to the authors, is mainly the merit of their specific morphology. The SERS activity depends also on the type of metal used and the increase in the following order Cu < <Au < Ag, wherein composite made of Cu/TiO_2-nanotube/Ti was less active than referential samples made from activated Cu bulk substrate. This observation could be attributed to an instantaneous oxidation of Cu clusters and particles on ATO surface, which result in quenching of SERS signals from the adsorbed molecules. Additionally, the highest substrate stability was observed for Au/TiO_2-nanotube/Ti composite.

Huang et al. [101] prepared SERS substrate consisting of Ag nanoparticles (NPs) deposited on patterned TiO_2 nanotube films through pulse-current (PC) electrodeposition. The resultant substrate exhibited particle-size-dependent as well as density-correlated UV–vis absorbance and SERS enhancement effect. It was shown that SERS signal of R6G molecules obtained on the tested sample was highly enhanced.

Differences in ATO and AAO matrices are not only in their morphology. Kudelski et al. compared SERS substrate behavior consisting of Ag nanoparticles deposited on AAO and ATO nanotubes [102] for pyridine and two various selected dyes. The recorded average SERS enhancement factors on Ag/TiO_2-n/Ti and Ag/Al_2O_3-n/Al composites was at least equal to the SERS enhancement factor on standard electrochemically roughened Ag substrates (i.e., about 10^6–10^7) or higher by a factor of 2. The authors proven that SERS spectra for examined molecules are distinctly different for those two composite substrates. They proposed that observed differences may be caused by specific interaction between the Ag nanoparticles and the oxides, especially different location of Fermi level in the Ag nanoparticles deposited on two tested nanostructured oxides, AAO and ATO. Moreover, the authors suggested that the critical influence on the shape of the spectrum (the position of individual SERS signals) and most probably on the "chemical" part of SERS enhancement is the electronic properties of the composites, not the specific surface area of substrate.

In other paper from the same group, Pisarek et al. reported [85] the influence of the size of the nanotubes/nanopores of AAO and ATO decorated by Ag nanoparticle as well as the structure of the porous oxide layers on the SERS enhancement factor (EF) for pyridine and mercaptobenzoic acid detection. It was shown that for the same amount of deposited Ag (0.02 mg/cm^2), the size of the nanopores significantly affects the EF, which reaches distinctly higher values at an optimal nanopore size (wherein the higher the EF, the smallest the nanotube diameter) than on a standard silver surface roughened by electrochemical cycling, i.e., EF > 10^6. Moreover it was shown that for Ag/Al_2O_3 NT/Al material the EF is significantly higher than that for Ti-based composite and an order of magnitude higher than that for commonly used in SERS measurements roughened silver. It was proven that porous alumina interact with Ag nano-deposits, hence affecting the electronic structure of the Ag-nanoparticles, which may be most useful in detecting small amounts of certain organic compounds.

Other researchers, Bao et al. [82] and Zhou et al. [83], combined advantages of using AAO matrix to fabricate ordered nanostructures with catalytic activity of TiO_2. Based on the photocatalytic properties of titanium dioxide, the authors have obtained promising self-cleaning effect of adsorbed molecules from composite SERS substrates made of Ag nanowires (made by AAO template-assisted fabrication method) coated with a TiO_2 layer. That composite was used as a SERS substrate as prepared as well as after cleaning realized by UV light irradiation. SERS effect was tested successfully for R6G (10^{-14} M) and methyl parathion (MP) (up to 10^{-6} M) on both, fresh and recycled samples [83]. The authors gave reasons why tested composite showed high SERS enhancement [83]:

1. Ag provided much higher SERS enhancement than Au and other metals.

2. Between collapsing nanowire tips, the gaps are created and they act as many "hot spots," where analyte molecules are positioned.

3. High aspect ratio (length/width) of Ag nanowires is a major factor to achieving high SERS enhancement.

5. Future of SERS substrate

The challenge that needs to be solved in order to use the SERS spectroscopy method on a wider scale than now is to develop a cheaper and repeatable method for the synthesis of uniformly distributed homogeneous nanostructures enabling the production of SERS substrates significantly enhancing the signal (>10^6) in a repeatable manner.

The ideal SERS substrate should be characterized by enhanced high electromagnetic field, not have a fluorescent background, and operate in a wide wavelength range and with unlimited laser power. In addition, it should have a long shelf life, be repeatable and uniform throughout its entire surface, be able to operate in different environments, and be available in different sizes. The future SERS substrates will be a commercially produced, cheap, and repeatable substrate, using new plasmonic materials (e.g., semiconductors, graphene), possible to use in portable Raman spectrometers. It is very likely that databases for the interpretation of standardized SERS spectra will be created in the future.

Taking into account all the requirements for an ideal SERS substrate, which are mentioned above, the anodic oxide-based nanocomposites as SERS substrate can become an efficient and popular product. Nowadays, the possibility of applying this material as a highly active SERS substrate is in the research phase, and the results are very promising. The prospective role of advanced anodic oxide based nanocomposites as SERS substrate can be attributed to the following features:

- No fluorescence and very low background, similar to the values recorded for quartz and silicon.

- The production of anodic oxides is relatively easy and fast process, so the cost of such production would be significantly lower than the cost of now commercially available SERS substrates.

- The structures obtained by anodization are characterized by a large specific surface area, repeatability, and the ability to control geometric parameters by means of manufacturing process parameters (e.g., anodization potential or process temperature), which ensure the repeatability of SERS analyses.

- It is possible to increase the production scale of anodic oxide-based SERS substrates relatively quickly and easily.

6. Conclusion

Despite the undoubted advantages and potentially large possibilities of the widespread use of SERS spectroscopy in many field, its progress has been hampered by the inability of scientists and industry to produce SERS substrates with high sensitivity, stability, and repeatability. Currently commercially available SERS substrates are characterized by obtaining enhancement of 10^5–10^7 and durability between several days and 6 months. The above limitations appear due to the relatively small control over the morphology of substrate structures at the nanoscale, which are responsible for the observed enhancement of SERS [103].

It is important to have the reproducible and homogeneous distribution of nanostructures over the entire surface of the SERS substrate, while obtaining large areas that meet these conditions. The next step is to find technology to transfer these results from the laboratory to the industrial scale providing some technological problems. The use of conventional nanostructures in fabrication methods becomes less attractive because they are relatively expensive, time-consuming, difficult, and not effective for areas larger than 1 mm^2. The use of anodic aluminum oxide template for the production of nanostructures, despite promising results at the nanoscale, is fraught with many problems arising when attempting to increase the scale of production, including a complicated, multi-stage, and difficult process for producing ultrathin membranes; ordering structures only in the domain area; and difficulties in obtaining repetitive systems of nanostructure. On the other hand, the outstanding array structure made by AAO template-assisted method has a promising practical application in SERS field, so this direction will be probably developed in the future. Also using the ATO for SERS substrate production seems to be a promising solution: the possibility of combining relatively ordered nanopores/nanotubes with its photocatalytic properties to utilize SERS substrates self-cleaning effect gives a big perspective of application on that material. In the future probably the composite SERS substrates based on AAO/ATO and Ag or Au will be commercially available.

Acknowledgements

This work was financed by the National Centre for Research and Development (Poland) under program LIDER IX (Nr LIDER/50/0199/L-9/17/NCBR/2018). M. Michalska-Domańska cordially acknowledges financial support from the Polish Ministry of Science and Higher Education (Scholarship for Young, Outstanding Researchers 2016–2019, agreement no. 1013/E-410/STYP/11/2016). The author thanks Dr. P. Nyga for the help in finding information about commercially available SERS substrates.

Author details

Marta Michalska-Domańska
Institute of Optoelectronics, Military University of Technology, Warsaw, Poland

*Address all correspondence to: marta.michalska@wat.edu.pl

References

[1] Douketis C, Haslett TL, Wang Z, Moskovits M, Iannotta S. Self-affine silver films and surface-enhanced Raman scattering: Linking spectroscopy to morphology. The Journal of Chemical Physics. 2000;**113**:11315-11323

[2] Fleischmann M, Hendra PJ, McQuillan AJ. Raman spectra of pyridine adsorbed at a silver electrode. Chemical Physics Letters. 1974;**26**(2):163-166

[3] Jeanmaire DL, Van Duyne RP. Surface Raman spectroelectrochemistry: Part I. heterocyclic, aromatic, and aliphatic amines adsorbed on the anodized silver electrode. Journal of Electroanalytical Chemistry and Interfacial Electrochemistry. 1977;**84**(1):1-20

[4] Tu Q, Chang C. Diagnostic applications of Raman spectroscopy. Nanomedicine: Nanotechnology, Biology, and Medicine. 2012;**8**:545-558

[5] Culha M, Cullum B, Lavrik N, Klutse CK. Surface-enhanced Raman scattering as an emerging characterization and detection technique, Hindawi Publishing Corporation. Journal of Nanotechnology. 2012;**2012**:971380. DOI: 10.1155/2012/971380

[6] Jha SK, Ekinci Y, Agio M, Loffler JF. Towards deep-UV surface-enhanced resonance Raman spectroscopy of explosives: Ultrasensitive, real-time and reproducible detection of TNT. Analyst. 2015;**140**:5671-5677. DOI: 10.1039/C4AN01719F

[7] Cao YWC, Jin R, Mirkin CA. Nanoparticles with Raman spectroscopic fingerprints for DNA and RNA detection. Science. 2002;**297**(5586):1536-1540. DOI: 10.1126/science.297.5586.1536

[8] Nariman B, Foley A, Houghton JM, Sun Y, Kim B. Multiplex detection of pancreatic cancer biomarkers using a SERS-based immunoassay. Nanotechnology. 2017;**28**(45):455101. DOI: 10.1088/1361-6528/aa8e8c

[9] Satpathy G, Chandra GK, Manikandan E, Roy Mahapatra D, Umapathy S. Pathogenic *Escherichia coli* (*E. coli*) detection through tuned nanoparticles enhancement study. Biotechnology Letters. 2020;**42**:853-863. DOI: 10.1007/s10529-020-02835-y

[10] Ben Jaber SS, Peveler WJ, Quesada Cabrera R, Sol CWO, Papakonstantinou I, Parkin IP. Sensitive and specific detection of explosives in solution and vapour by surface-enhanced Raman spectroscopy on silver nanocubes. Nanoscale. 2017;**9**:16459-16466. DOI: 10.1039/C7NR05057G

[11] Shanthil M, Fathima H, Thomas KG. Cost-effective plasmonic platforms: Glass capillaries decorated with Ag@SiO_2 nanoparticles on inner walls as SERS substrates. ACS Applied Materials & Interfaces. 2017;**9**:19470-19477. DOI: 10.1021/acsami.6b12478

[12] Zhang C, Wang K, Han D, Pang Q. Surface enhanced Raman scattering (SERS) spectra of trinitrotoluene in silver colloids prepared by microwave heating method. Spectrochimica Acta Part A. 2014;**122**:387-391. DOI: 10.1016/j.saa.2013.11.066

[13] Hamad S, Podagatlapalli GK, Mohiddon MA, Soma VR. Cost effective nanostructured copper substrates prepared with ultrafast laser pulses for explosives detection using surface enhanced Raman scattering. Applied Physics Letters. 2014;**104**:263104. DOI: 10.1063/1.4885763

[14] Lu H, Zhang H, Yu X, Zeng S, Yong K-T, Ho H-P. Seed-mediated plasmon-driven regrowth of silver nanodecahedrons (NDs). Plasmonics. 2011;7(1):167-173

[15] Coluccio ML, Das G, Mecarini F, Gentile F, Pujia A, Bava L, et al. Silver-based surface enhanced Raman scattering (SERS) substrate fabrication using nanolithography and site selective electroless deposition. Microelectronic Engineering. 2009;**86**:1085-1088

[16] Bao L-L, Mahurin SM, Liang C-D, Dai S. Study of silver films over silica beads as a surface-enhanced Raman scattering (SERS) substrate for detection of benzoic acid. Journal of Raman Spectroscopy. 2003;**34**(5):394-398

[17] Pilot R, Signorini R, Durante C, Orian L, Bhamidipati M, Fabris L. A review on surface-enhanced Raman scattering. Biosensors. 2019;**9**:57. DOI: 10.3390/bios9020057

[18] Wang Z, Zong S, Wu L, Zhu D, Cui Y. SERS-activated platforms for immunoassay: Probes, encoding methods, and applications. Chemical Reviews. 2017;**117**(12):7910-7963

[19] Hasna K, Antony A, Puigdollers J, Kumar KR, Jayaraj MK. Fabrication of cost-effective, highly reproducible large area arrays of nanotriangular pillars for surface enhanced Raman scattering substrates. Nano Research. 2016;**9**(10):3075-3083

[20] Zhang CP, Yi PY, Peng LF, Lai XM, Chen J, Huang MZ, et al. Continuous fabrication of nanostructure arrays for flexible surface enhanced Raman scattering substrate. Scientific Reports. 2017;**7**:39814-39822

[21] Maher RC. SERS Hot Spots. In: Kumar CSSR, editor. Raman Spectroscopy for Nanomaterials Characterization. Berlin, Heidelberg: Springer; 2012

[22] Pacaud M, Hervé-Aubert K, Soucé M, Makki A, Bonnier F, Fahmi A, et al. One-step synthesis of gold nanoflowers of tunable size and absorption wavelength in the red & deep red range for SERS spectroscopy. Spectrochimica Acta Part A: Molecular and Biomolecular Spectroscopy. 2020;**225**:117502. DOI: 10.1016/j.saa.2019.117502

[23] Wan Y. Modulation of localized surface plasmon resonance of the titanium nitride nanoparticle array based on graphene. Optik. 2020;**200**:163420. DOI: 10.1016/j.ijleo.2019.163420

[24] Araujo TP, Quiroz J, Barbosa ECM, Camargo PHC. Understanding plasmonic catalysis with controlled nanomaterials based on catalytic and plasmonic metals. Current Opinion in Colloid & Interface Science. 2019;**39**:110-122. DOI: 10.1016/j.cocis.2019.01.014

[25] Mejía-Salazar JR, Oliveira ON Jr. Plasmonic biosensing. Chemical Reviews. 2018;**118**:10617-10625. DOI: 10.1021/acs.chemrev.8b00359

[26] Zhu C, Xu Q. Amorphous materials for enhanced localized surface plasmon resonances. Chemistry, an Asian Journal. 2018;**13**(7):730-739. DOI: 10.1002/asia.201701722

[27] Agrawal A, Cho SH, Zandi O, Ghosh S, Johns RW, Milliron DJ. Localized surface plasmon resonance in semiconductor nanocrystals. Chemical Reviews. 2018;**118**(6):3121-3207. DOI: 10.1021/acs.chemrev.7b00613

[28] Csáki A, Stranik O, Fritzsche W. Localized surface plasmon resonance based biosensing. Expert Review of Molecular Diagnostics.

2018;**18**(3):279-296. DOI: 10.1080/14737159.2018.1440208

[29] Huang Z, Meng G, Huang Q, Chen B, Zhu C, Zhang Z. Large-area Ag nanorod array substrates for SERS: AAO template-assisted fabrication, functionalization, and application in detection PCBs. Journal of Raman Specroscopy. 2013;**44**:240-246

[30] Chettiar UK, Nyga P, Thoreson MD, Kildishev AV, Drachev VP, Shalaev VM. FDTD modeling of realistic semicontinuous metal films. Applied Physics B: Lasers and Optics. 2010;**100**(1):159-168. DOI: 10.1007/s00340-010-3985-y

[31] Wang AX, Kong X. Review of recent progress of plasmonic materials and nano-structures for surface-enhanced Raman scattering. Materials. 2015;**8**:3024-3052

[32] Bang D, Chang YW, Park J, Lee T, Park J, Yeo J-S, et al. One-step electrochemical fabrication of vertically self-organized silver nanograss. Journal of Materials Chemistry A. 2013;**1**:4851-4857

[33] Lan H, Ding Y. Nanoimprint lithography. In: Wang M, editor. Lithography. IntechOpen; 2010. DOI: 10.5772/8189. Available from: https://www.intechopen.com/books/lithography/nanoimprint-lithography

[34] Vieu C, Carcenac F, Pépin A, Chen Y, Mejias M, Lebib A, et al. Electron beam lithography: Resolution limits and applications. Applied Surface Science. 2000;**164**(1-4):111-117

[35] Lushi K, Nanxi D, Guofeng T, Shengli Q, Dezhen W. Highly enhanced Raman scattering with good reproducibility observed on a flexible PI nanofabric substrate decorated by silver nanoparticles with controlled size. Applied Surface Science. 2020;**511**:145443

[36] Shi J, You T, Gao Y, Liang X, Li C, Yin P. Large-scale preparation of flexible and reusable surface-enhanced Raman scattering platform based on electrospinning AgNPs/PCL nanofiber membrane. RSC Advances. 2017;**7**(75):47373-47379

[37] Chen X, Lin H, Xu T, Lai K, Han X, Linm M. Cellulose nanofibers coated with silver nanoparticles as a flexible nanocomposite for measurement of flusilazole residues in oolong tea by surface-enhanced Raman spectroscopy. Food Chemistry. 2020;**315**:126276

[38] Parnsubsakul A, Ngoensawat U, Wutikhun T, Sukmanee T, Sapcharoenkun C, Pienpinijtham P, et al. Silver nanoparticle/bacterial nanocellulose paper composites for paste-and-read SERS detection of pesticides on fruit surfaces. Carbohydrate Polymers. 2020;**235**:115956

[39] Lu J, Song Y, Lei F, Du X, Huo Y, Xu S, et al. Electric field-modulated surface enhanced Raman spectroscopy by PVDF/Ag hybrid. Scientific Reports. 2020;**10**:5269

[40] Liou P, Nayigizik FX, Kong F, Mustapha A, Lin M. Cellulose nanofibers coated with silver nanoparticles as a SERS platform for detection of pesticides in apples. Carbohydrate Polymers. 2017;**157**:643-650

[41] Gao R, Song X, Zhan C, Weng C, Cheng S, Guo K, et al. Light trapping induced flexible wrinkled nanocone SERS substrate for highly sensitive explosive detection. Sensors and Actuators B: Chemical. 2020;**314**:128081

[42] Stępniowski WJ, Norek M, Michalska-Domańska M, Bombalska A, Nowak-Stępniowska A, Kwaśny M, et al. Fabrication of anodic aluminum oxide with incorporated chromate

ions. Applied Surface Science. 2012;**259**:324-330

[43] Macak JM, Tsuchiya H, Schmuki P. High-aspect-ratio TiO_2 nanotubes by anodization of titanium. Angewandte Chemie, International Edition. 2005;**44**(14):2100-2102

[44] Macak JM, Tsuchiya H, Taveira L, Aldabergerova S, Schmuki P. Smooth anodic TiO_2 nanotubes. Angewandte Chemie, International Edition. 2005;**44**(45):7463-7465

[45] Mika K, Socha RP, Nyga P, Wiercigroch E, Małek K, Jarosz M, et al. Electrochemical synthesis and characterization of dark nanoporous zinc oxide films. Electrochimica Acta. 2019;**305**:349-359

[46] Zaraska L, Gawlak K, Wiercigroch E, Malek K, Koziel M, Andrzejczuk M, et al. The effect of anodizing potential and annealing conditions on the morphology, composition and photoelectrochemical activity of porous anodic tin oxide films. Electrochimica Acta. 2019;**319**:18-30

[47] Syrek K, Zaraska L, Zych M, Sulka GD. The effect of anodization conditions on the morphology of porous tungsten oxide layers formed in aqueous solution. Journal of Electroanalytical Chemistry. 2018;**829**:106-115

[48] Chilimoniuk P, Michalska-Domańska M, Czujko T. Formation of nanoporous mixed aluminum-iron oxides by self-organized anodizing of $FeAl_3$ intermetallic alloy. Materials. 2019;**12**(14):2299

[49] Łazińska M, Durejko T, Michalska-Domańska M, Bojar Z. Characterization of titanium oxide formed on biomedical Ti_6Al_7Nb alloy fabricated by laser engineered net shaping (LENS). In: Proceedings Euro PM2017: International Power Metallurgy Congress and Exhibition 2017. Milan, Italy: Additive Manufacturing; 2017. p. 140833

[50] Stępniowski WJ, Choi J, Yoo H, Oh K, Michalska-Domańska M, Chilimoniuk P, et al. Anodization of FeAl intermetallic alloy for nanoporous mixed aluminum-iron oxide. Journal of Electroanalytical Chemistry. 2016;**771**:37-44

[51] Stępniowski WJ, Choi J, Yoo H, Michalska-Domańska M, Chilimoniuk P, Czujko T. Quantitative fast Fourier transform based arrangement analysis of nanoporous anodic oxide formed by self-organized anodization of FeAl intermetallic alloy. Materials Letters. 2016;**164**:176-179

[52] Stępniowski WJ, Paliwoda D, Chen Z, Landskron K, Misiołek WZ. Hard anodization of copper in potassium carbonate aqueous solution. Materials Letters. 2019;**252**:182-185

[53] Stepniowski WJ, Yoo H, Choi J, Chilimoniuk P, Karczewski K, Czujko T. Investigation of oxide nanowires growth on copper via passivation in NaOH aqueous solution. Surface and Interface Analysis. 2019;**14**:15-18

[54] Stępniowski WJ, Yoo H, Choi J, Norek M, Jozwik P, Misiolek WZ. Fabrication and characterization of oxide nano-needles formed by copper passivation in sodium hydroxide solution. Thin Solid Films. 2019;**671**:111-119

[55] Stępniowski WJ, Moneta M, Karczewski K, Michalska-Domańska M, Czujko T, Mol JMC, et al. Fabrication of copper nanowires via electrodeposition in anodic aluminum oxide templates formed by combined hard anodizing and electrochemical barrier layer thinning. Journal of Electroanalytical Chemistry. 2018;**15**:59-66

[56] Stępniowski WJ, Salerno M. Fabrication of nanowires and nanotubes by anodic alumina template-assisted electrodeposition. In: Manufacturing

Nanostructures. UK: One Central Press; 2014

[57] Stępniowski WJ, Bojar Z. Nanoporous anodic aluminum oxide: Fabrication, characterization, and applications. In: Handbook of Nanoelectrochemistry: Electrochemical Synthesis Methods, Properties, and Characterization Techniques. Switzerland: Springer; 2016. pp. 593-646

[58] Abrahami ST, de Kok JMM, Terryn H, Mol JMC. Towards Cr(VI)-free anodization of aluminum alloys for aerospace adhesive bonding applications: A review. Frontiers of Chemical Science and Engineering. 2017;**11**(3):465-482

[59] Abrahami ST, Hauffman T, de Kok JMM, Mol JMC, Terryn H. XPS analysis of the surface chemistry and interfacial bonding of barrier-type Cr(VI)-free anodic oxides. Journal of Physical Chemistry C. 2015;**119**(34):19967-19975

[60] Kulkarni M, Mazare A, Schmuki P, Iglic A. Influence of anodization parameters on morphology of TiO_2 nanostructured surfaces. Advanced Materials Letters. 2016;7(1):23-28

[61] Naghizadeh M, Abdizadeh H, Golobostanfard MR. Effect of fluoride concentration and water content on morphology of titania nanotubes in ethylene glycol solution. Advanced Materials Research. 2013;**829**:907-911

[62] Michalska-Domańska M, Nyga P, Czerwiński M. Ethanol-based electrolyte for nanotubular anodic TiO_2 formation. Corrosion Science. 2018;**134**:99-102

[63] Zhang S, Li Y, Xu P, Liang K. Effect of anodization parameters on the surface morphology and photoelectrochemical properties of TiO_2 nanotubes. International Journal of Electrochemical Science. 2017;**12**(11):10714-10725

[64] Michalska-Domańska M, Stępniowski W, Salerno M. Effect of inter-electrode separation in the fabrication of nanoporous alumina by anodization. Journal of Electroanalytical Chemistry. 2018;**823**:47-53

[65] Michalska-Domanska M, Stępniowski WJ, Jaroszewicz LR. Characterization of nanopores arrangement of anodic alumina layers synthesized on low-(AA1050) and high-purity aluminum by two-step anodizing in sulfuric acid with addition of ethylene glycol at low temperature. Journal of Porous Materials. 2017;**24**(3):779-786

[66] Stępniowski WJ, Moneta M, Norek M, Michalska-Domańska M, Scarpellini A, Salerno M. The influence of electrolyte composition on the growth of nanoporous anodic alumina. Electrochimica Acta. 2016;**211**:453-460

[67] Stępniowski W, Norek M, Budner B, Michalska-Domańska M, Nowak-Stępniowska A, Mostek A, et al. In-situ electrochemical doping of nanoporous anodic aluminum oxide with indigo carmine organic dye. Thin Solid Films. 2016;**598**:60-64

[68] Stępniowski WJ, Florkiewicz W, Michalska-Domańska M, Norek M, Czujko T. A comparative study of electrochemical barrier layer thinning for anodic aluminum oxide grown on technical purity aluminum. Journal of Electroanalytical Chemistry. 2015;**741**:80-86

[69] Stępniowski WJ, Forbot D, Norek M, Michalska-Domańska M, Król A. The impact of electrolyte's viscosity on the formation of nanoporous anodic aluminum oxide. Electrochimica Acta. 2014;**133**:57-64

[70] Stępniowski WJ, Norek M, Michalska-Domańska M, Forbot D, Król A. Study on the correlation between criterion number derived from Rayleigh-Bénard convective cells and arrangement of nanoporous anodic aluminum oxide. Materials Letters. 2014;**125**:124-127

[71] Michalska-Domańska M, Norek M, Stępniowski WJ, Budner B. Fabrication of high quality anodic aluminum oxide (AAO) on low purity aluminum – A comparative study with the AAO produced on high purity aluminum. Electrochimica Acta. 2013;**105**:424-432

[72] Wang Y, Wang Y, Wang H, Wang X, Cong M, Xu W, et al. Hierarchical ultrathin alumina membrane for the fabrication of unique nanodot arrays. Nanotechnology. 2016;**27**:025302. DOI: 10.1088/0957-4484/27/2/025302

[73] Nguyen TT, Ung TDT, Nguyen QL. Square-inch 2D-arrays of Au nanodisks fabricated by sputtering Au onto anodic aluminum oxide templates for SERS applications. Advances in Natural Sciences: Nanoscience and Nanotechnology. 2016;**7**:045017

[74] Liu D, Wang Q, Hu J, Chen L. Size correlation of optical and SERS properties for highly ordered Au nanocone arrays with sub-100 nm feature size. Journal of Optics. 2016;**18**:085006

[75] Li ZP, Xu ZM, Qu XP, Wang SB, Peng J, Mei LH. Fabrication of nanopore and nanoparticle arrays with high aspect ratio AAO masks. Nanotechnology. 2017;**28**:095301

[76] Kim Y-T, Schilling J, Schweizer SL, Wehrspohn RB. Morphology dependence on surface-enhanced Raman scattering using gold nanorod arrays consisting of agglomerated nanoparticles. Plasmonics. 2017;**12**:203-208. DOI: 10.1007/s11468-016-0250-1

[77] Feng Y, Kim K-D, Nemitz CA, Kim P, Pfadler T, Gerigk M, et al. Uniform large-area free-standing silver nanowire arrays on transparent conducting substrates. Journal of the Electrochemical Society. 2016;**163**(8):D447-D452

[78] Hao Q, Huang H, Fan X, Hou X, Yin Y, Li W, et al. Facile design of ultra-thin anodic aluminum oxide membranes for the fabrication of plasmonic nanoarrays. Nanotechnology. 2017;**28**:105301

[79] Liu K, Yu Z, Zhu X, Zhang S, Zou F, Zhu Y. A universal surface enhanced Raman spectroscopy (SERS)-active graphene cathode for lithium–air batteries. RSC Advances. 2016;**6**:102272. DOI: 10.1039/c6ra23331g

[80] Zhao W, Wu Y, Liu X, Xu Y, Wang S, Xu Z. The fabrication of polymer-nanocone-based 3D Au nanoparticle array and its SERS performance. Applied Physics A: Materials Science & Processing. 2017;**123**:45. DOI: 10.1007/s00339-016-0665-8

[81] Zhao WN, Liu XG, Xu YB, Wang SB, Sun TY, Liu SS, et al. Polymer nanopillar array with Au nanoparticle inlays as a flexible and transparent SERS substrate. RSC Advances. 2016;**6**:35527-35531

[82] Bao Z, Wu Y. TiO_2 thin layer coated Ag nanoarrays complex for surface-enhanced Raman scattering substrate. Key Engineering Materials. 2013;**562-565**:1037-1042. DOI: 10.4028/www.scientific.net/kem.562-565.37

[83] Zhou Y, Chen J, Zhang L, Yang L. Multifunctional TiO_2-coated Ag nanowire arrays as recyclable SERS substrates for the detection of organic pollutants. European Journal of Inorganic Chemistry. 2012;**2012**:3176-3182

[84] Zhang C-y, Lu Y, Zhao B, Hao Y-w, Liu Y-q. Facile fabrication of Ag

dendrite-integrated anodic aluminum oxide membrane as effective three-dimensional SERS substrate. Applied Surface Science. 2016;**377**:167-173

[85] Pisarek M, Holdynski M, Roguska A, Kudelski A, Janik-Czachor M. TiO_2 and Al_2O_3 nanoporous oxide layers decorated with silver nanoparticles—Active substrates for SERS measurements. Journal of Solid State Electrochemistry. 2014;**18**:3099-3109

[86] Chen H, Ohodnicki P, Baltrus JP, Holcomb G, Tylczak J, Henry D. High-temperature stability of silver nanoparticles geometrically confined in the nanoscale pore channels of anodized aluminum oxide for SERS in harsh environments. RSC Advances. 2016;**6**:86930

[87] Lim L-K, Ng B-K. Large scale AAO nano-fiber substrate for SERS application. In: Photonics North (PN). Quebec City, QC; 2016. pp.1-2. DOI: 10.1109/PN.2016.7537899

[88] Dan Y, Zhong C, Zhu H, Wang J. Highly ordered Au-decorated Ag nanorod arrays as an ultrasensitive and reusable substrate for surface enhanced Raman scattering. Colloids and Surfaces, A: Physicochemical and Engineering Aspects. 2019;**560**:360-365

[89] Zhong C, Dan Y, Zhang P, Wang J. Self-assembly urchin-like Au-NSs arrays and application as surface-enhanced Raman scattering substrates. Materials Letters. 2019;**234**:125-128

[90] Yang L, Ruan W, Jiang X, Zhao B, Xu W, Lombardi JR. Contribution of ZnO to charge-transfer induced surface-enhanced Raman scattering in Au/ZnO/PATP assembly. Journal of Physical Chemistry C. 2009;**113**:117-120

[91] Yang L, Jiang X, Ruan W, Yang J, Zhao B, Xu W, et al. Charge-transfer-induced surface-enhanced Raman scattering on Ag-TiO_2 nanocomposites. Journal of Physical Chemistry C. 2009;**113**:16226-16231. DOI: 10.1021/jp903600r

[92] Yang L, Jiang X, Ruan W, Zhao B, Xu W, Lombardi JR. Observation of enhanced Raman scattering for molecules adsorbed on TiO_2 nanoparticles: Charge-transfer contribution. Journal of Physical Chemistry C. 2008;**112**:20095-20098

[93] Jakubowicz J, Koper JK, Adamek G, Połomska M, Wolak J. Silver Nano-trees deposited in the pores of anodically oxidized titanium and Ti scaffold. International Journal of Electrochemical Science. 2015;**10**:4165-4172

[94] Belich NA, Grigor'eva AV, Petukhov DI, Sidorov AV, Gol'dt AE, Gudilin EA. Immobilization of nanostructured metal silver at the surface of anodic titanium dioxide for the creation of composites with the surface plasmon resonance. Nanotechnologies in Russia. 2015;**10**(5-6):345-352

[95] Witkowska E, Szymborski T, Waluk J, Michota-Kamińska A. The platform for testing of chemicals and microorganisms via surface enhanced Raman spectroscopy and method of its preparation; 2014. p. 409210

[96] Kamińska A, Sivanesan A, Witkowska E, Gołąb J, Winiarska M, Nowis D, et al. Detection of DNA mutations using novel surface-enhanced Raman spectroscopy (SERS) diagnostic platform. Journal of Chemistry and Chemical Engineering. 2013;**7**:199-208

[97] Sun Y, Yang L, Liao F, Dang Q, Shao M. Parameter optimization for Ag-coated TiO_2 nanotube arrays as recyclable SERS substrates. Applied Surface Science. 2018;**443**:613-618

[98] Wen S, Su Y, Wu R, Zhou S, Min Q, Fan G-C, et al. Plasmonic Au nanostar

Raman probes coupling with highly ordered TiO_2/Au nanotube arrays as the reliable SERS sensing platform for chronic myeloid leukemia drug evaluation. Biosensors and Bioelectronics. 2018;**117**:260-266

[99] Ling Y, Zhuo Y, Huang L, Mao D. Using Ag-embedded TiO_2 nanotubes array as recyclable SERS substrate. Applied Surface Science. 2016;**388**:169-173

[100] Roguska A, Kudelski A, Pisarek M, Opara M, Janik-Czachor M. Surface-enhanced Raman scattering (SERS) activity of Ag, Au and Cu nanoclusters on TiO_2-nanotubes/Ti substrate. Applied Surface Science. 2011;**257**:8182-8189

[101] Huang Y, Sun L, Xie K, Lai Y, Liu B, Ren B, et al. SERS study of Ag nanoparticles electrodeposited on patterned TiO_2 nanotube films. Journal of Raman Specroscopy. 2011;**42**:986-991. DOI: 10.1002/jrs.2830

[102] Kudelski A, Pisarek M, Roguska A, Hołdyński M, Janik-Czachor M. Surface-enhanced Raman scattering investigations on silver nanoparticles deposited on alumina and titania nanotubes: Influence of the substrate material on surface-enhanced Raman scattering activity of Ag nanoparticles. Journal of Raman Specroscopy. 2012;**43**:1360-1366

[103] Banholzer MJ, Millstone JE, Qin L, Mirkin CA. Rationally designed nanostructures for surface-enhanced Raman spectroscopy. Chemical Society Reviews. 2008;**37**:885-897

Chapter 5

An Experimental Investigation of Al_2O_3-40% TiO_2 Powder Amalgamated via Atmospheric Plasma Spray Coating onto SS316 Substrate and Parameter Optimization Using TLBO Algorithm

Thankam Sreekumar Rajesh

Abstract

SS316 is a commercial stainless steel. MTBF (Mean Time Between Failure) of SS 316 wear prone areas can be effectively increased by ceramic coating. The coating thickness, surface roughness, coating microhardness, abrasion rate, and coating porosity decides the quality and durability in ceramic coating. The current research work explains an experimental investigation to optimize the Atmosphere Plasma Spray process input parameters of Al_2O_3-40%TiO_2 ceramic coatings. Three-level L_{18} Orthogonal Array (OA) design of Experiments (DoE) is used to conduct the current work. The main input parameters considered in the current study are nozzle distance, substrate speed, arc current, carrier gas flow, and coating powder flow rate. The output parameters considered are coating thickness, surface roughness, coating microhardness, abrasion rate, and percentage of porosity. Mathematical models are generated for individual output parameters. AHP (Analytical Hierarchy Process) is effectively used to find out weights for individual output parameters treating them as objective functions, and a combined objective function is generated.

Keywords: atmospheric plasma spray (APS) coating, SS316, teaching learning based optimization (TLBO), Al_2O_3-40%TiO_2

1. Introduction

To achieve increased reliability and performance of damage prone industrial-related components, surface engineering is hugely now applied using large field of new technologies. The quest for higher efficiency and productivity across the entire spectrum of manufacturing and engineering industries has ensured that most of the machine components are subjected to highly harsh environments during routine operation [1, 2]. The high deterioration of parts and their ultimate failure has been traced to material damage bought in by hostile environments like high relative motion between mating services, corrosive media, extreme temperatures, and cyclic stresses. As a result of the above, the concept of applying engineered surfaces capable of combating the high degradation phenomena like wear, corrosion, and fatigue to improve component performance, reliability, and life cycle has gained high acceptance in last one decade. To act as a last line of defense, a proactive coating deposited to act as a perfect barrier between the initial surface of the component and the aggressive environment that is exposed during routine operation is now globally acknowledged as an attractive and effective solution to significantly reduce damage.

316 stainless steel is widely used in chemical/petrochemical industry, food processing, pharmaceutical equipment manufacturing, medical devices fabrication, in potable water/wastewater treatment/marine applications and architectures. SS316 composition consists of Chromium 10–14%, Nickel 2–3%, carbon 16–18%, and Molybdenum. In general, the addition of about 2% Molybdenum in SS316 stainless steel provides excellent level of resistance against pitting corrosion and stress corrosion cracking especially in a saline environment compared to SS304. Abrasion and wear resistance is the type of damage which needs to be improved during its application in industrial environment. Abrasion resistance capabilities of SS316 can be highly enhanced by providing a coating of Al_2O_3-40%TiO_2 on an SS316 steel surface by means of atmospheric plasma spraying.

Hence Al_2O_3-40%TiO_2 coating on SS316 surface makes it an ideal combination to combat today's highly drastic and hostile industrial environments where assets are sweating to achieve target of productivity and yield which is always more than 100% of the rated capacity. Even though there is a high level of interest and expectations regarding the fundamentals and development of atmospheric plasma arc spraying, there is a huge gap of reliable as well as dependable models that correlates final engineering properties of coatings such as surface roughness, microhardness, porosity, etc. with variations in critical input process parameters and geometry of deposition process [3].

In the present work, Al_2O_3-40% TiO_2 is coated on SS316 substrates and all the desired critical output parameters are measured and recorded for further analysis.

2. Experimentation details

Justification for the usage of Al_2O_3-40% TiO_2 amalgamated powder.

Al_2O_3-40% TiO_2 amalgamated powder is used for all the experiments conducted during this research work. The use of Al_2O_3-40% TiO_2 justified as it helps to decrease the melting temperature of amalgamated composite powder and hence the final coating exhibits very low porosity % and enhanced fracture toughness compared to 97/3 and 87/13 Al_2O_3 and TiO_2 combinations. The coating generated with Al_2O_3-40% TiO_2 amalgamated powder also possesses high dielectric strength, enhanced wear and heat resistance [4].

2.1 Development of experimental plan

As per the literature review, the following process parameters play a deciding role in the Al_2O_3-TiO_2atmospheric plasma spray process on various substrates.

Major input parameters of a plasma arc spray are below:

1. Carrier gas flow rate (Ar: 20–45 L/min, H_2: 8–14 L/min)
2. Powder feed rate (19–56 g/min)
3. Electric power input/Arc current (250–600 Amps)
4. Substrate rpm
5. Spray distance (70–200 mm)
6. Flame temperature (14000–16,000°C)
7. Composition of working gas
8. Fuel gas to oxygen ratio
9. Powder particle size
10. Powder morphology
11. Pre-spray sand blasting particle size (20/24/60 μm)
12. Bond coat material type (Ni Cr 80/20, 60/40, 70/30)
13. Post spray treatment
14. Substrate temperature (40–200°C)
15. Spray gun Coolant type (Chilled water or normal water, air)
16. Mixing combination of Al_2O_3/TiO_2 (13% Wt TiO_2, 40% Wt TiO_2, 20%Wt TiO_2)
17. Gas injection angle and parameters related with gun
18. Nozzle diameter (6, 6.5, 7, 7.5, 8 mm, GP, GH, GE, G Profiles)
19. Sealing after coating
20. Use of device for coating (Manipulator/robot/manual)
21. RPM of the job, while coating (100–500 RPM)
22. Distance between the substrate and the nozzle (75–125 mm)

The major output parameters are listed below:

1. Porosity %
2. Coating thickness

No	Parameter	Low level	Middle level	High level
1	Spray distance of gun, mm	75	100	125
2	Carrier gas flow, lit./min.	20	30	50
3	Powder flow rate, g/min	25	35	50
4	RPM of the substrate	150	250	350
5	Arc current, A	350	400	500

Table 1.
Input parameters with three levels.

3. Microhardness
4. Abrasion resistance (abrasion rate)
5. Surface roughness
6. Oxidation on surface
7. Microhardness
8. Abrasion resistance, abrasion rate
9. Bonding strength
10. Wear resistance
11. Dielectric strength

After considering the critical requirements of the industry, review of facilities available and thorough literature survey, the following input parameters with three levels are considered for experiments as shown as **Table 1**.

Similarly, the output parameters selected are: coating thickness, surface roughness, microhardness, abrasion rate, and porosity %.

3. Design of experiments

Orthogonal array experimental design proposed by Taguchi can be efficiently used to examine the effect of different input parameters on the critical performance characteristics linked output parameters through compact set of experiments [5]. In-depth understanding of the process, including the minimum, maximum, and current value of the parameter is required to decide the set of input parameters that are highly affecting a process as well as the levels at which these parameters should be varied [6].

After thoroughly understanding the number of input parameters and the number of levels a proper compact can be finalized using the array selector as shown in **Table 2**, by looking at the column and row corresponding to the number of parameters and number of levels.

An L18 orthogonal array is used to carry out all the experiments as shown in **Table 3** considering the five input parameters with three levels. **Table 4** shows the complete experiment plan.

Number of levels		Number of parameters																	
		2	3	4	5	6	7	8	9	10	11	12	13	14	15	16	17	18	19
	2	L4	L4	L8	L8	L8	L8	L12	L12	L12	L12	L16	L16	L16	L16	L32	L32	L32	L32
	3	L9	L9	L9	L18	L18	L18	L18	L27	L27	L27	L27	L27	L36	L36	L36	L36	L36	L36
	4	L16	L16	L16	L16	L32	L32	L32	L32	L32									
	5	L25	L25	L25	L25	L25	L50	L50	L50	L50	L50	L50							

Table 2.
Array selector for DoE.

Experiment	P1	P2	P3	P4	P5
1	1	1	1	1	1
2	1	2	2	2	2
3	1	3	3	3	3
4	2	1	1	2	2
5	2	2	2	3	3
6	2	3	3	1	1
7	3	1	2	1	3
8	3	2	3	2	1
9	3	3	1	3	2
10	1	1	3	3	2
11	1	2	1	1	3
12	1	3	2	2	1
13	2	1	2	3	1
14	2	2	3	1	2
15	2	3	1	2	3
16	3	1	3	2	3
17	3	2	1	3	1
18	3	3	2	1	2

Table 3.
L18 orthogonal array.

SN	Spray distance mm	Substrate rpm	Arc current A	Carrier gas flow L/min	Powder flow rate g/min
1	75	150	300	20	25
2	75	250	400	30	35
3	75	350	500	40	50
4	125	150	300	30	35
5	125	250	400	40	50
6	125	350	500	20	25
7	175	150	400	20	50
8	175	250	500	30	25
9	175	350	300	40	35
10	75	150	500	40	35
11	75	250	300	20	50
12	75	350	400	30	25
13	125	150	400	40	25
14	125	250	500	20	35
15	125	350	300	30	50
16	175	150	500	30	50
17	175	250	300	40	25
18	175	350	400	20	35

Table 4.
Complete plan of experiments as per L18 OA.

4. Experimental steps

Al_2O_3-40%TiO_2 powder from H C Starck, USA is utilized for coating for all the set of experiments. SS316 substrates are prepared with 27.5 mm diameter and 3 mm thickness plates and pieces of size 75 × 25 ×12 mm are prepared for coating so as to conduct the Abrasion test [7].

Each experiment is carried out with three substrate samples. The substrate samples are assembled in one specially fabricated cartridge [8]. To avoid non-uniformity in thickness, the substrate samples are ground to achieve relatively good finish. Later, sand blasting is carried out on the surface of all substrate samples to ensure proper removal of oxides and other impurities [9, 10]. For sand blasting, fused alumina of grit size 60 μm from Carborandum Universal is used [11, 12]. Al_2O_3-40% TiO_2 powder is deposited on the substrates by using a controlled atmospheric plasma stray system of Metco USA through an SG 100 model Plasma Gun. To ensure the removal of moisture, the amalgamated powder is preheated before the plasma coating process up to 110°C [13].

As pre the L18 orthogonal array DoE, the experiments are conducted. The coating facility arrangement is shown in **Figure 1**. The parameters which are kept constant during the experiment are

Spray nozzle	GP, diameter: 5.43 mm
Grind blasting pressure	2 kg/cm^2
Substrate exposure to gun	30 s
Primary gas pressure	100 Psi
Secondary gas pressure	80 Psi

The substrate samples are cleaned after the coating with ethanol and properly dried to eliminate accumulation of moisture [14]. As per ASTM B499-92014 [15], coating thickness is measured. An ultrasonic thickness gauge is used for this purpose. Using Mitutoyo surface roughness tester SV-C3100, surface roughness is measured as per ASTM D127 2013 [16]. The surface testing probe is applied on the coating surface for a length of 15 mm with a pitch of 0.001 mm. The scanning speed is kept constant at 2.0 mm/s. Microhardness is measured as per ASTM B 578-872,015 [17] by Metatech MVH-1. To measure porosity as well as microhardness, the samples are sectioned by wire cutting as well as slow speed grinding.

Later, molds are prepared using Bain mount-3 molding machine supplied by Chennai Matco. The substrate surfaces on the mold are polished with emery papers

Figure 1.
Coating facility arrangement.

ranging from 150 grit up to 2000 grit so as to get scale free and super finished surface. The porosity is measured using an Image Analysis Software as shown in **Figure 2**. The JPG images of the coatings are taken by using an electronic microscope OLIMBUS GX51 as per ASTM E2109-012014 [18]. A sample image of the sectioned coating surface is shown as **Figure 3**. Abrasion rate is measured on Abrasion test rig TR-50 as per ASTM G65 2000 [19] with 1000 g. as pre-load. **Figure 4** shows one of the abrasion test in progress. This test method covers laboratory procedures for determining the resistance of coating materials to scratching abrasion by means of the dry sand/rubber wheel test. Abrasion test results are reported as weight loss in grams for 20 revolutions of the wheel for a particular substrate. Materials of higher abrasion rate and high weight loss show a lower abrasion resistance.

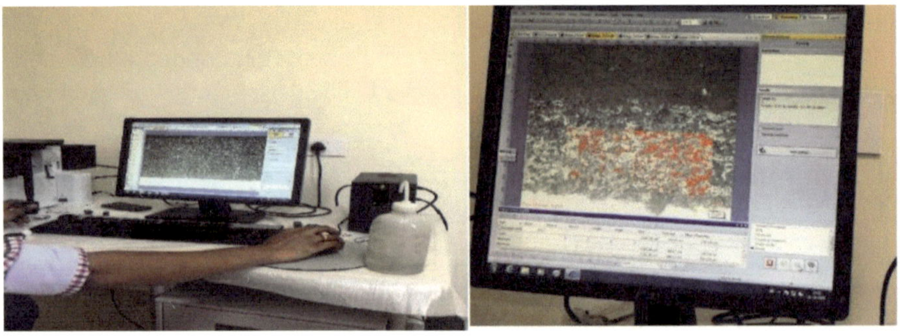

Figure 2.
Measurement of porosity in progress.

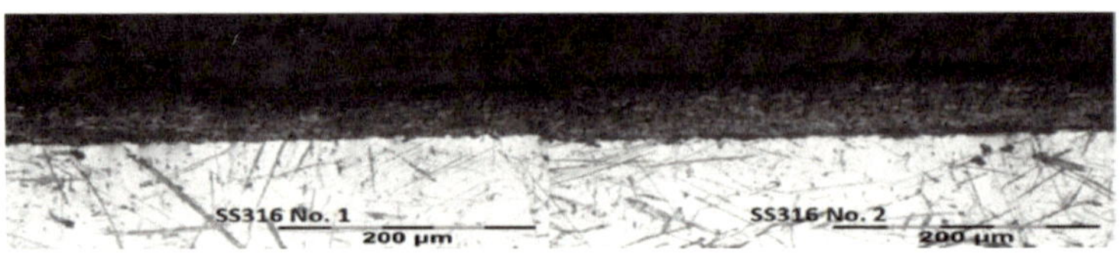

Figure 3.
Sample photograph of the sectioned coated surface.

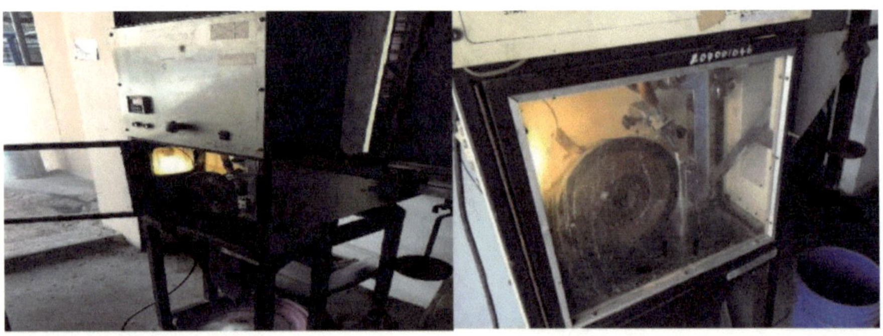

Figure 4.
Abrasion test in progress.

4.1 Details of coating output parameters

The measured values of coating thickness, surface roughness, microhardness, abrasion rate, and porosity % are given in **Table 5**.

To generate mathematical models for each of the output parameters, excel data analysis is used [18]. The mathematical models generated for all the output parameters are shown below as Eqs. (1) and (2).

$$\begin{aligned} \text{T, Thickness } (\mu\text{m}) = {} & 1934.148 - 27.5317 \times D - 4.39986 \times N - 2.41271 \times A \\ & + 23.17571 \times G + 12.19609 \times P + 0.017778 \times D \times N \\ & + 0.009069 \times D \times A + 0.265778 \times D \times G + 0.287119 \times D \times P \\ & + 0.00872 \times N \times A - 0.02146 \times N \times G + 0.015764 \times N \times P \\ & + 0.019442 \times A \times G - 0.00704 \times A \times P - 1.40365 \times G \times P \end{aligned} \tag{1}$$

$$\begin{aligned} \text{R, Roughness } (\mu\text{m}) = {} & 13.669392 + 0.029373 \times D - 0.007159 \times N - 0.005758 \times A \\ & - 0.52605 \times G - 0.093101 \times P - 0.000006 \times D \times N \\ & - 0.000184 \times D \times A + 0.001429 \times D \times G + 0.00035 \times D \\ & \times P + 0.000029 \times N \times A - 0.000099 \times N \times G + 0.000056 \\ & \times N \times P + 0.000789 \times A \times G - 0.000001 \times A \times P \\ & + 0.000907 \times G \times P \end{aligned} \tag{2}$$

No	Mean coating thickness, μm	Mean surface roughness, μm	Mean microhardness HV	Mean abrasion rate, g	Mean porosity, %
1	343.33	5.2	953	0.1602	14.2243
2	186.67	4.22	929	0.1322	11.5634
3	226.67	5.46	906	0.1998	13.2112
4	170	4.48	749	0.1023	18.0264
5	433.33	4.54	802	0.0601	11.4133
6	236.67	5.1	953	0.0855	9.5114
7	286.67	4.33	766	0.0973	14.7045
8	206.67	4.36	716	0.1022	18.9231
9	376.67	5.36	821	0.135	15.0654
10	366.67	4.72	766	0.0355	9.0012
11	516.67	4.35	841	0.0546	36.2234
12	600	4.74	841	0.0788	17.2133
13	416.67	4.36	821	0.0786	19.4231
14	476.67	4.39	802	0.0688	19.7324
15	323.33	4.42	821	0.1001	25.3381
16	346.67	3.95	802	0.1001	22.1099
17	203.33	5.43	766	0.1255	10.5224
18	283.33	5.03	784	0.0979	21.3422

Table 5.
The values of measured output parameters–SS316.

$$\begin{aligned}\text{H, Microhardness (HV)} = {} & 2866.016 - 14.81975 \times D - 5.535376 \times N - 2.302965 \\ & \times A - 6.870915 \times G - 11.56905 \times P + 0.010359 \times D \\ & \times N - 0.002454 \times D \times A + 0.207755 \times D \times G \\ & + 0.180859 \times D \times P + 0.010923 \times N \times A - 0.016017 \\ & \times N \times G + 0.030724 \times N \times P + 0.016899 \times A \times G \\ & + 0.00575 \times A \times P - 0.54882 \times G \times P\end{aligned} \tag{3}$$

$$\begin{aligned}\text{A, Abrasion rate (g)} = {} & 0.84561453 + 0.00023070 \times D - 0.00160816 \times N \\ & - 0.00127953 \times A - 0.00654575 \times G - 0.01238197 \times P \\ & - 0.00000033 \times D \times N + 0.00000114 \times D \times A \\ & - 0.00001169 \times D \times G - 0.00000214 \times D \times P \\ & + 0.00000106 \times N \times A + 0.00002805 \times N \times G \\ & + 0.00001562 \times N \times P + 0.00000362 \times A \times G \\ & + 0.00002127 \times A \times P - 0.00001584 \times G \times P\end{aligned} \tag{4}$$

$$\begin{aligned}\text{P, Porosity (\%)} = {} & 43.36101115 - 0.55405377 \times D + 0.03082400 \times N - 0.18845218 \\ & \times A + 1.40724984 \times G + 0.74786458 \times P + 0.00002005 \times D \\ & \times N + 0.00120547 \times D \times A + 0.00284447 \times D \times G \\ & - 0.00062184 \times D \times P - 0.00004973 \times N \times A - 0.00199498 \\ & \times N \times G + 0.00226109 \times N \times P + 0.00060184 \times A \times G \\ & + 0.00091991 \times A \times P - 0.04473134 \times G \times P\end{aligned} \tag{5}$$

where, D = spray distance; N = substrate rpm; A = arc current; G = carrier gas flow rate; P = powder flow rate.

The values of roughness, abrasion rate and porosity are considered non-beneficial and the values of thickness and hardness are considered as beneficial.

4.2 Confirmation experiments

For conducting confirmation tests, three trial samples of SS316 are used. The random values for all the input parameters, in between the maximum and

Spray distance mm	Substrate rpm	Arc current A	Carrier gas flow L/min	Powder flow rate g/min	Coating thickness µm	Predicted values µm	% variation
150	300	450	35	30	278.33	306.8094	10.2322
160	225	325	25	30	206.33	218.2186	5.7619
80	190	425	35	40	322.67	296.4082	8.1388

Table 6.
Measured and predicted values – Thickness, SS316.

Spray distance mm	Substrate rpm	Arc current A	Carrier gas flow L/min	Powder flow rate g/min	Coating roughness µm	Predicted values µm	% Variation
100	200	350	25	40	4.23	4.476552	5.8286
150	300	450	35	30	4.38	5.264112	20.1852
160	225	325	25	30	5.02	5.577542	11.1064

Table 7.
Measured and predicted values – Roughness, SS316.

Spray distance mm	Substrate rpm	Arc current A	Carrier gas flow rate L/min	Powder flow rate g/min	Hardness HV	Predicted values HV	% Variation
150	300	450	35	30	894	914.1079	2.2492
160	225	325	25	30	578	624.3352	8.0165
80	190	425	35	40	642	770.9902	20.0919

Table 8.
Measured and predicted values – Microhardness, SS316.

Spray distance mm	Substrate rpm	Arc current A	Carrier gas flow L/min	Powder flow rate g/min	Abrasion rate g	Predicted values g	% Variation
115	180	325	25	30	0.0911	0.088834	2.4868
120	215	425	35	40	0.0987	0.118253	19.8107
85	295	495	25	45	0.1265	0.125752	0.5911

Table 9.
Measured and predicted values – Abrasion rate, SS316.

Spray distance mm	Substrate rpm	Arc current A	Carrier gas flow rate L/min	Powder flow rate g/min	Porosity %	Predicted values %	% Variation
80	170	315	25	30	21.33	18.52049	13.1802
95	190	425	35	40	20.01	18.41184	7.9868
140	200	480	35	45	9.56	10.53198	10.1532

Table 10.
Measured and predicted values – Porosity %, SS316.

minimum levels are taken to conduct the confirmation tests. The predicted values using the proposed model along with the measured output parameters values for three samples are given in **Tables 6–10**. The percentage variation between the actual and predicted values are also shown in the tables.

5. SN analysis

To determine the effect each variable has on the output, the signal-to-noise ratio, or the SN number, needs to be calculated for each experiment conducted. In the equations below, y_i is the mean value and s_i is the variance. y_i is the value of the performance characteristic for a given experiment. More details about SN analysis is given in https://google site sn analysis design of experiments.

SN ratio values for coating thickness on SS316 substrate are calculated for each parameter and level. The values are tabulated as shown in **Table 11**.

Spray distance has a significant effect on the coating thickness. Carrier gas flow is the next dominant parameter in the case of coating thickness. Similarly, SN ratio values are calculated for surface roughness for each parameter and level for all the output parameters as shown in **Table 12**. In the case of surface roughness, carrier gas flow has a significant effect. Substrate rpm is the next influencing parameter as per R value. The spray distance has a significant effect on the microhardness as shown in **Table 13**. The next dominant parameter in the case of microhardness is rpm. For coating abrasion rate, spray distance is the most dominating parameter and then comes the rpm as shown in **Table 14**. In the case of coating porosity % as

	Spray distance	Substrate rpm	Arc current	Carrier gas flow rate	Powder flow rate
Level 1	50.73	49.83	49.58	50.7	49.77
Level 2	50.16	49.74	50.72	48.81	49.24
Lvel 3	48.83	50.15	49.42	50.2	50.71
Δ R	1.90	0.41	1.29	1.89	1.47
Rank	1	5	4	2	3

Table 11.
SN ratio matrix and ΔR values of coating thickness, SS316.

	Spray distance	Substrate rpm	Arc current	Carrier gas flow rate	Powder flow rate
Level 1	13.56	13.05	13.72	13.47	13.71
Level 2	13.14	13.12	13.12	12.78	13.41
Level 3	13.46	13.99	13.32	13.9	13.03
Δ R	0.42	0.94	0.6	1.12	0.68
Rank	5	2	4	1	3

Table 12.
SN ratio matrix and ΔR values of coating roughness, SS316.

	Spray distance	Substrate rpm	Arc current	Carrier gas flow rate	Powder flow rate
Level 1	58.79	58.14	58.3	58.55	58.45
Level 2	58.3	58.13	58.3	58.14	58.13
Level 3	57.79	58.61	58.28	58.19	58.3
Δ R	1.01	0.48	0.03	0.42	0.32
Rank	1	2	5	3	4

Table 13.
SN ratio matrix and ΔR values of microhardness, SS316.

	Spray distance	Substrate rpm	Arc Current	Carrier gas flow l/min	Powder flow rate g/min
Level 1	−20.63	−21.17	−19.4	−21.03	−19.88
Level 2	−21.82	21.39	−21.1	−19.87	−21.2
Level 3	−19.27	−19.16	−21.23	−20.82	−20.65
Δ R	2.55	2.23	1.84	1.16	1.32
Rank	1	2	3	5	4

Table 14.
SN ratio matrix and ΔR values of coating abrasion rate, SS316.

	Spray distance	Substrate rpm	Arc current	Carrier gas flow rate	Powder flow rate
Level 1	23.62	23.88	25.24	24.94	23.18
Level 2	24.26	24.25	23.8	25.27	23.59
Level 3	24.39	24.15	23.23	22.07	25.51
Δ R	0.77	0.37	2.01	3.2	2.33
Rank	4	5	3	1	2

Table 15.
SN ratio matrix and ΔR values of porosity %, SS316.

shown in **Table 15**, the most dominating parameter is carrier gas flow. The next parameter which effects the most is powder flow rate.

6. Application of teaching learning based optimization (TLBO)

An advanced optimization method, known as Teaching-Learning-Based Optimization (TLBO) is applied, to determine the best values of input parameters to obtain global optimum output parameters. TLBO is applied individually to each of the developed mathematical models given by Eqs. (1)–(5). Rao et al., proposed TLBO, which is based on the effect of influence of a teacher on the output of learners in a class. Teaching-learning ability of teacher and learners in a class room is mimicked in this algorithm [20, 21]. There are two modes of learning in this algorithm, interacting with other learners (known as learner phase) and through teacher (known as teacher phase).One of the attractive features of this algorithm is its algorithm-specific parameter-less concept. The algorithm is widely preferred among researchers due to its simplicity and its ability to provide the global optimum solutions in comparatively less number of function evaluations. Further details about the TLBO algorithm can be found at https://sites.google.com/site/tlborao.

To execute TLBO algorithm for the optimisation of individual objective functions, a population size of 10 and 100 number of iterations with 30 independent runs is considered. The global optimum values for individual objective functions of T (Coating thickness), R (Surface roughness), H (Microhardness), Ab (Abrasion rate) and Po (Porosity %) obtained after applying TLBO are given in **Table 16**.

Values obtained by applying the TLBO algorithm for the individual objective functions of T (Thickness), R (Roughness), H (Microhardness), Ab (Abrasion rate

Objective function (Output parameter)	Optimum values of input parameters					Value of the objective function
	Spray distance	Substrate rpm	Arc current	Carrier gas flow	Powder flow rate	
Coating thickness	175	350	500	40	50	1068.8 μm
Surface roughness	75	350	300	40	50	1.1503 μm
Microhardness	175	350	500	40	50	1396 HV
Abrasion rate	75.34	150	300	40	50	0.0073 g
Porosity %	75	150	500	20	25	1.4935%

Table 16.
Optimized output parameter values obtained by applying TLBO.

and Po (Porosity %) are 1068.8 μm, 1.1503 μm, 1396 HV, 0.0073 g, and 1.4935% respectively. For each of the output parameters, the corresponding values of process input parameters are also given in **Table 16**. It can be observed from **Table 16** that the optimum values of input parameters for getting optimum value of a particular objective (i.e., output parameter) are not the same for the other objectives. In real industrial situations, it is required to find the set of optimum values of input parameters that satisfies all the objectives simultaneously. Hence, the problem becomes a multi-objective problem with the ranges of the input parameters as constraints. In the current work, a combined objective function is formed considering all the five objectives simultaneously. This is called a priori approach of solving the multi-objective optimization problems.

7. Formation of combined objective function

A pirori approach is used by forming a combined objective function, involving all the three objectives and this function is solved by applying TLBO algorithm for the given ranges of the input parameters.

$$Z_{max} = \omega_T * \frac{T}{T_{max}} + \omega_H * \frac{H}{H_{max}} - \omega_{Ab} * \frac{Ab}{Ab_{min}} - \omega_{Po} * \frac{Po}{Po_{min}} - \omega_R * \frac{R}{R_{min}} \quad (6)$$

The normalized weights of each output parameters are calculated using AHP (Arithmetic Hierarchy Method) [22] and these are $\omega_T = 0.4027$, $\omega_R = 0.0694$ $\omega_H = 0.2595$ $\omega_{Ab} = 0.1342$ and $\omega_{Po} = 0.1342$. The weights are applied to combined objective function as given below in Eq. (7). For more details about AHP method pl. refer https://google site AHP Saaty.

$$\begin{aligned} Z_{max} = {} & [(0.4027/1068.8) \times (1934.148 - 27.5317 \times D - 4.39986 \times N \\ & -2.41271 \times A + 23.17571 \times G + 12.19609 \times P + 0.017778 \times D \times N \\ & + 0.009069 \times D \times A + 0.265778 \times D \times G + 0.287119 \times D \times P \\ & + 0.00872 \times N \times A - 0.02146 \times N \times G + 0.015764 \times N \times P \\ & + 0.019442 \times A \times G - 0.00704 \times A \times P - 1.40365 \times G \times P)] \\ & + [(0.2595/1396) \times (2866.016 - 14.81975 \times D - 5.535376 \times N \\ & -2.302965 \times A - 6.870915 \times G - 11.56905 \times P + 0.010359 \times D \\ & \times N - 0.002454 \times D \times A + 0.207755 \times D \times G + 0.180859 \times D \\ & \times P + 0.010923 \times N \times A - 0.016017 \times N \times G + 0.030724 \times N \\ & \times P + 0.016899 \times A \times G + 0.00575 \times A \times P - 0.54882 \times G \times P)] \\ & - (0.1342/0.0073) \\ & \times (0.84561453 + 0.00023070 \times D - 0.00160816 \times N \\ & -0.00127953 \times A - 0.00654575 \times G - 0.01238197 \times P \\ & - 0.00000033 \times D \times N + 0.00000114 \times D \times A - 0.00001169 \\ & \times D \times G - 0.00000214 \times D \times P + 0.00000106 \times N \times A \\ & + 0.00002805 \times N \times G + 0.00001562 \times N \times P + 0.00000362 \\ & \times A \times G + 0.00002127 \times A \times P - 0.00001584 \times G \times P)] \\ & - [(0.1342/1.4935) \times (43.36101115 - 0.55405377 \times D \\ & +0.03082400 \times N - 0.18845218 \times A + 1.40724984 \times G \\ & + 0.74786458 \times P + 0.00002005 \times D \times N + 0.00120547 \times D \\ & \times A + 0.00284447 \times D \times G - 0.00062184 \times D \times P \\ & - 0.00004973 \times N \times A - 0.00199498 \times N \times G + 0.00226109 \\ & \times N \times P + 0.00060184 \times A \times G + 0.00091991 \times A \times P \\ & - 0.04473134 \times G \times P)] \\ & - [(0.0694/1.1503) \times (13.669392 + 0.029373 \times D - 0.007159 \times N \end{aligned}$$

$$-0.005758 \times A - 0.52605 \times G - 0.093101 \times P - 0.000006 \times D \times N - 0.000184 \times D \times A + 0.001429 \times D \times G + 0.00035 \times D \times P + 0.000029 \times N \times A - 0.000099 \times N \times G + 0.000056 \times N \times P + 0.000789 \times A \times G - 0.000001 \times A \times P + 0.000907 \times G \times P)] \tag{7}$$

TLBO algorithm can be again applied on the combined objective function and after n number of iterations with independent runs to achieve the global optimum value of coefficient Z_{max} and the corresponding values of the optimum input parameters can be arrived.

8. Conclusion

In the field of atmospheric plasma coating with Al_2O_3-40%TiO_2, mathematical modeling and its optimization is very rare. Mathematical models are generated in the present work using regression analysis for all the output parameters in terms of input parameters. The mathematical models developed will work as effective tools for manufactures to predict the effect of input parameters on output parameters within the considered ranges. Based on this model, they can take decisions and hence costly trials can be avoided to a very large extent. Confirmation tests are also carried out in the present work for each of the output parameters. The confirmation tests have given near about the same values compared to the predicted values and the percentage of error is negligible.

The optimization is effectively carried out using teaching-learning-based optimization (TLBO) algorithm for each output parameter individually. A combined objective function is generated and this combined objective function can be again optimized using TLBO algorithm to get global optimum values of input parameters considering all the output parameters simultaneously. The TLBO algorithm has proved its effectiveness and simplicity in solving the multi-objective optimization problems. The AHP method is applied to decide the weights for the individual objective functions in the combined objective function in a systematic way and it takes into account the preferences of the decision maker. It is also concluded that a change in the weights of the individual objective functions in the combined objective function may give different sets of optimum values of input parameters.

Author details

Thankam Sreekumar Rajesh
SVNIT, Surat, Gujarat, India

*Address all correspondence to: rajeshtsreekumar@gmail.com

References

[1] Ctibor P, Stengl V, Pis I, Zahoranova T, Nehasil V. Plasma sprayed TiO2: The influence of an electric supply on relations among stoichiometry, surface state and photo catalytic decomposition of acetone. Ceramics International. 2012;**38**: 3453-3458

[2] Datta S, Pratihar DK, Bandyopadhyay PP. Modelling of plasma spray coating process using statistical regression analysis. International Journal of Advanced Manufacturing Technology. 2013;**65**: 967-980

[3] Forghani SM, Ghazali MJ, Muchtar A, Daud AR. Mechanical properties of plasma sprayed nanostructured TiO2 coatings on mild steel. Ceramics International. 2014;**40**: 7049-7056

[4] Ghazali MJ, Forghani SM, Hassanuddin N, Muchtar A, Daud AR. Comparative wear study of plasma sprayed TiO2 and Al2O3–TiO2 on mild steels. Integrated Tribology International. 2016;**93**:681-686

[5] Hazra S, Bandyopadhyay PP. Scratch induced failure of plasma sprayed alumina based coatings. Materials and Design. 2012;**35**:243-250

[6] Morks MF, Akimoto K. The role of nozzle diameter on the microstructure and abrasion wear resistance of plasma sprayed Al2O3/TiO2 composite coatings. Journal of Manufacturing Processes. 2008;**10**:1-5

[7] Mishra NK, Mishra SB, Kumar R. Oxidation resistance of low-velocity oxy fuel-sprayed Al2O3-13TiO2 coating on nickel-based superalloys at 800 °C. Surface and Coating Technology. 2014; **260**:23-27

[8] Paneerselvam R. Design and Analysis of Experiments. New Delhi: PHI Learning Pvt. Ltd.; 2012

[9] Ramachandran CS, Balasubramanian V, Ananthapadmanabhan PV, Viswabaskaran V. Influence of intermixed interfacial layers on the thermal cycling behavior of atmospheric plasma sprayed lanthanum zirconate based coatings. Ceramics International. 2012;**38**:4081-4096

[10] Rico A, Poza P, Rodriguez J. High temperature tribological behaviour of nanostructured and conventional plasma sprayed alumina-titania coatings. Vacuum. 2012;**888**:1-6

[11] Saravanan P, Selvarajan V, Joshi SV, Sundararajan G. Experimental design and performance analysis of alumina coatings deposited by a detonation spray process. Journal of Applied Physics. 2001;**34**:131-140

[12] Sathish S, Geetha M, Aruna ST, Balaji N, Rajam KS, Asokamani R. Sliding wear behaviour of plasma sprayed nanoceramic coatings for biomedical applications. Wear. 2011; **271**:934-941

[13] Sathish S, Geetha M, Aruna ST, Balaji N, Rajam KS, Asokamani R. Studies on plasma sprayed bilayered ceramic coatings for bio-medical Ti-13Nb-13Zr ally. Ceramics International. 2011;**37**:1333-1339

[14] Christel P, Lech P, Murel B, Peirre C. Design of experiments in thermal spraying. A review. Surface and coatings Technology. 2008;**202**: 4483-4490

[15] ASTM B499-09. Standard Test Method for Measurement of Coating

Thicknesses by the Magnetic Method. 2014. Available from: www.astm.org

[16] ASTM D7127. Standard Test Method for Measurement of Surface Roughness of Abrasive Blast Cleaned Metal Surfaces Using a Portable Stylus Instrument. 2013. Available from: www.astm.org

[17] ASTM B578-87. Standard Test Method for Microhardness of Electroplated Coatings. 2015. Available from: www.astm.org

[18] ASTM E2109-0. Standard Test Methods for Determining Area Percentage Porosity in Thermal Sprayed Coatings. 2014. Available from: www.astm.org

[19] ASTM G65. Standard Test Method for Measuring Abrasion Using the Dry Sand/Rubber Wheel Apparatus. 2000. Available from: www.astm.org

[20] Rao RV. Teaching Learning Based Optimization Algorithm and its Engineering Applications. Switzerland: Springer International Publishing; 2016

[21] Rao RV. Review of applications of TLBO algorithm and a tutorial for beginners to solve the unconstrained and constrained optimization problem, Decision Science Letters; 2015

[22] Saaty TL. Fundamentals of the Analytic Hierarchy Process. Pittsburg: RWS Publications; 2000

Section 4

2D-2D Metal - Non-Metal Matrix

Chapter 6

LDH Ternary Nanocomposites: g-C_3N_4 Intercalated ZnO\Mg-Al for Superior Photocatalytic Activity towards Dye Degradation

Kandasamy Bhuvaneswari, Thangavelu Pazhanivel, Govindasamy Palanisamy and Ganapathi Bharathi

Abstract

Photocatalytic dye degradation has received more attention as an affordable and effective way to treat the dye polluted water. In the present chapter, we are going to discuss; (i) the preparation and photophysical characterization of g-C_3N_4 intercalated ZnO\Mg-Al LDH, a novel ternary nanocomposite, and (ii) its visible light photocatalytic degradation activity against the methylene blue dye. LDHs are 2D materials composed of "brucite-like" cationic layers where an inclusion of trivalent cations presents an overall positive charge to the nanosheets. g-C_3N_4 is one of the organic semiconductor photocatalyst which active for several types of reactions such as CO_2 reduction, water splitting, and degradation because of its stable, non-toxic, and earth-abundant nature. Mainly, the development of numerous 2D g-C_3N_4 nanosheets has been extensively used in the field of photocatalyst. By the combination heterojunction with 2D/2D interface can effectively improve the photocatalytic activity. The nitrogen-rich g-C_3N_4 intercalated ZnO\Mg-Al LDH ternary nanocomposite formation could follow the direct dye degradation process and results enhance the visible light absorption. The enhanced photocatalytic activity is mainly due to the improved charge separation rate and high number of photogenerated electrons. The large number of photogenerated electrons and high charge separation efficiency are effectively influence the dye degradation efficiency.

Keywords: layered double hydroxides, graphitic carbon nitride, photocatalytic activity, methylene-blue, ternary nanocomposites, visible light, dye degradation

1. Introduction

1.1 Water pollution

The industrial revolution could not avoid its effects on increasing environmental pollution, which pose a life threat to living beings. On the other hand, the increase of population rises the corresponding needs, which in turn result in the increased release of pollutants. The toxic substances from farmhouses, municipalities,

IntechOpen

pesticides, and factories are the major sources of water pollution. Organic dyes are one of the major groups of pollutants which are released from textile industrial wastewater. The dye effluent contaminates the surface and groundwater, thereby, making it unfit for drinking and other daily usages. Polluted drinking water can cause serious cariogenic effects on human and other living beings.

The effective handling of increasing environmental pollution is a major challenge for the sustainable progress of modern civilization. With a lack of waste management measures, there is an urgent need in finding efficient ways to treat and decompose the pollutants. Water is a "universal solvent," it can dissolve more substances than any other liquid on earth. It is because of this substantial property, water dissolves most of the pollutants and thus be polluted easily. Quality drinking water is a fundamental right to every human being and most of the countries do not provide drinking water in the WHO standards. Water pollution not only affects the human being, but also every living organism, as there is nothing without water.

Organic dyes used in many industries such as textiles, furniture, chemical, paint, food, and cosmetic industries are the major water pollutants. The organic dyes possess color owing to the following reasons; (i) the dye molecules absorb light in the visible region of the electromagnetic spectrum (400–700 nm), (ii) they have a conjugated structure, i.e. a structure with alternating single and double bonds, (iii) the molecule dye have at least one color bearing chromophore group, and (iv) exhibit resonance of electrons, which is a stabilizing force in organic compounds [1]. The removal of dye molecules is a challenging process because of the enormous variety of functional groups in dissimilar dyes and their different properties. Many techniques like electrochemical coagulation, reverse osmosis, nano-filtration, photocatalytic degradation, adsorption using activated materials etc., are used for the removal of dye from wastewater. Among the various types of approaches adsorption and photocatalytic degradation of chemically stable organic pollutants occupy a prominent place, due to some of the obvious advantages such as cost-effectiveness, simplicity of operation besides great efficiency.

1.2 Photocatalysis

Photocatalysis is a process, which accelerates a photoreaction in the presence of a photocatalyst. Photocatalysis, as a fresh, cheap, environmentally friendly "green" process, offers great potential for environmental protection and energy exchange. The organic pollutants can be effectively decomposed by the semiconductor-based photocatalysts under light irradiation with the photon energies equal or higher to the bandgaps of the photocatalysts. In recent years, the photocatalytic reaction has received increasing attention for environmental applications such as air purification, hazardous material remediation, water disinfection, and water purification. The versatility of the photocatalytic process, for example, photocatalytic degradation of dyes and photoelectrocatalytic reduction of CO_2 into hydrocarbon compounds in aqueous semiconductor suspensions, greatly attracted the scientists to work in the field of photocatalysis. The pioneering work of photo-electrochemical water splitting on TiO_2 electrode reported by Fujishima and Honda in 1972, has been the initiative in the field of photocatalysis. In this way, the semiconductor based photocatalysis has grown as an ideal green chemistry tool in dealing with the globally concerned energy shortage and environmental pollution issues. In general, a photocatalytic reaction consists of three simple steps; (i) The semiconductor photocatalysts absorb incident photons whose energy ($h\nu$) is equal to or more than its bandgap (E_g), resulting in the generation of electron-hole pairs, (ii) The photogenerated charge electrons and holes are separated and transferred to the surface of photocatalysts, and (iii) The photogenerated electrons and holes contribute in catalytic reactions by forming superoxide and hydroxide radicals which react with dye molecules [2].

The detailed photocatalytic mechanism was shown in **Figure 1**. Several efforts have been conveyed through a variety of materials and methods and it is true, that each report put forward some scientific development to its ancestors. The photodegradation is one of the cost-effective and easy-to-implement methods, and the materials studied include TiO_2, SnO_2, ZnO, CuO, and WO_3 along with their heterostructures and organic/inorganic composites [3–7].

The photocatalytic decomposition of pollutants in the real-time application for water sanitization requires the use of non-toxic, cheap as well as reproducible resources. The conventionally used wide bandgap (SCs) with limited light responding range, which can only absorb UV light (λ <380 nm), seriously confines the photocatalytic efficiencies. Therefore, it has become a significant problem to develop the photocatalytic SCs with a visible light response for practical applications. Besides, another major task in photocatalysis is the increase in the charge separation efficiency of the photocatalyst and the corresponding photocatalytic efficiency. The separation of the electron–hole pairs can increase the efficiency of photocatalysts. Transition metal oxides (TiO_2, ZnO_2, SnO_2, etc.) have lower photocatalytic efficiency since its wide bandgap and high recombination rate of photogenerated electron-hole pairs. To overcome this difficulty, the development of hetero-nanostructures could offer an enhancement in the photocatalytic efficiency and can act as a better photocatalyst which can degrade various kinds of persistent organic pollutants.

Among the numerous photocatalytic materials, ZnO occupied the reasonable research area owing to its whole beneficial characteristics over other materials [8, 9]. Even though, when it comes to commercial developments, the robustness of ZnO needs further developments [10]. Such as, the trapping state (including interstitial and missing atoms/vacancy defect) bolstered loss of excitons, which is basic in oxide-based semiconductors, should be tended to appropriately [11]. One plausibility of accomplishing this is, utilize a better surfactant/capping molecule to passivate the surface traps, which overwhelmingly trigger the charge carrier recombination. On the other hand, such passivation has the opportunity to acting as a barrier for hinders the association between the dye pollutant and active material, which will likewise bring low efficiencies. The development of a ZnO based hybrid photocatalyst comprising of a composite material with suitable band structure would be a better choice towards the concealment of charge carrier recombination and consequent improvement in the photocatalytic dye degradation process [12, 13].

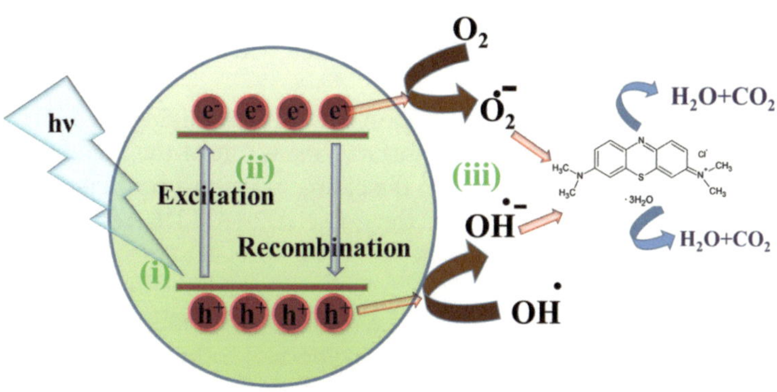

Figure 1.
Schematic illustration of photocatalytic dye degradation.

1.3 Graphitic carbon nitride (g-C_3N_4)

Graphitic carbon nitride (g-C_3N_4), is a two-dimensional metal-free conjugated crystalline sheet material with a bandgap energy of 2.7 eV, which has concerned exceptional research enthusiasm because of its environmental friendly nature, attractive electronic structure, low-cost excellent thermal and chemical stabilities [14–18]. The conduction and valence band boundaries of g-C_3N_4, exist at −1.12 and + 1.6 eV, making it active under visible light as an efficient photocatalyst [19–22]. Even though, its implication has drawbacks such as faster recombination of the electron-hole pairs, and agglomeration in most solvents caused by the strong van der Waals attractions between sp^2 carbon atoms [23]. 2D g-C_3N_4, nanosheets have much attention because of their enlarged specific surface area, improved electron–phonon interaction, and enhanced electron mobility along the in-plane direction [24]. Although some developments have been attained, the light-harvesting ability and quantum efficiency of these modified g-C_3N_4 systems are still poor.

For these reasons, various protocols such as surface modification, doping with metal or nonmetal elements and co-polymerization have been actively employed to enhance the photocatalytic performance of g-C_3N_4. It has high nitrogen content compared to other N-carbon materials, which is capable of creating more active reaction sites that would increase the electron donor/acceptor characteristics. Even after several decades and extensive investigations on several materials, a robust combination of materials and method is still required to vanish away the environment threatening organic pollutants.

1.4 Layered double hydroxides

Layered double hydroxides (LDH), a new class of lamellar metal hydroxide materials, consist of positively-charged hydrotalcite-like layers with carbonate ions and water molecules in the interlayer galleries [25–27]. Due to the two dimensional (2D) layered structure, LDH has a high explicit surface area, which can help quick ion transfer [26, 28–30]. Dvininov et al. prepared the SnO_2/Mg-Al LDH coupling through the thermal treatment, which demonstrated good photocatalytic activity for methylene blue degradation [31]. It was made conceivable by the oxygen reduction and progressive creation of hydroxyl radicals, which are accountable for the degradation. Seftel et al. synthesized Ti incorporated Mg-Al LDH solid which shows better photocatalytic activity due to the isolation of small TiO_2 nanoparticles on the LDH surface [32]. Kingshuk Dutta et al. prepared ZnO\Zn-Al LDH nanostructure by hydrothermal method using Al substrate as a template for developing different compositions and morphologies and the author demonstrated the degradation of Congo red dye [33]. Therefore, the LDH is a better candidate to be hybridized with ZnO which will enhance the catalytic activity of photocatalysts.

In any case, to build up a superior photocatalyst, hybridizing the LDH with a material having high conductivity and surface area is one of the hopeful approaches, which can further improve the charge transport proficiency of LDH-composite. Xiaoya Yuan et al. prepared the g-C_3N_4\Zn-Al LDH composites through a simple in situ crystallization technique and the as-prepared composite exhibited improved photodecolorization of MB.

In the present work, we have prepared a ternary nanocomposite of g-C_3N_4 intercalated ZnO\Mg-Al LDH through a hydrothermal technique and studied its photocatalytic activity against the MB dye degradation. The ZnO is attached on the surface also interlayers of the LDHs, and ZnO\Mg-Al LDH are distributed over the surface of g-C_3N_4 nanosheets. The nitrogen-rich ternary composite formation resulted in the enhancement of visible light absorption and improved charge separation to result in the enhanced photocatalytic degradation activity towards the MB dye.

2. Experimental section

2.1 Preparation of g-C_3N_4 nanosheets

The g-C_3N_4 was prepared by a thermal condensation method using melamine as a precursor. 5 g of melamine was kept in an alumina crucible and thermally treated at 550°C for 3 h in a furnace. The obtained agglomerate residues are ground into fine powder and subjected to hydrochloric acid treatment for 12 h to obtain the g-C_3N_4 nanosheets. The suspension was centrifuged to separate the residual of g-C_3N_4 nanosheets. The obtained precipitate product was heated at 60°C for overnight to attain the light yellow colored powder of g-C_3N_4 nanosheets.

2.2 Preparation of Mg-Al LDH

In a typical synthesis procedure, Mg-Al LDH was prepared by a facile hydrothermal method. Firstly, 0.05 M of aluminum nitrate and 0.03 M of magnesium chloride were dissolved into 20 ml DDW separately under vigorous magnetic stirring for 10 min. Subsequently, 0.04 M of urea were dissolved into the 10 ml DDW and stirred for 30 min. After that, the precursor and urea solutions were mixed and 0.2 M of NaOH was added to the above solution mixture until the pH of the suspension was reached 12. The entire solution was transferred into a 100 ml Teflon lined stainless-steel autoclave, followed by heating in an oven under 180°C for 24 h. After the reaction was complete, the autoclave was cooled to room temperature. Finally, the sample was centrifuged and washed with DDW water and dried at 80°C for overnight to obtain the final product.

2.3 Preparation of ZnO nanoparticles

ZnO nanoparticles were prepared by the hydrothermal method. In this process, 0.2 M of $ZnCl_2$ were dissolved in 100 ml of DDW, and 0.2 M of NaOH were dissolved in 20 ml of DDW separately under constant stirring. After 10 min stirring the above-mentioned solutions were mixed together, and transferred into a 100 ml Teflon liner stainless-steel autoclave, followed by heating in an oven under 180°C

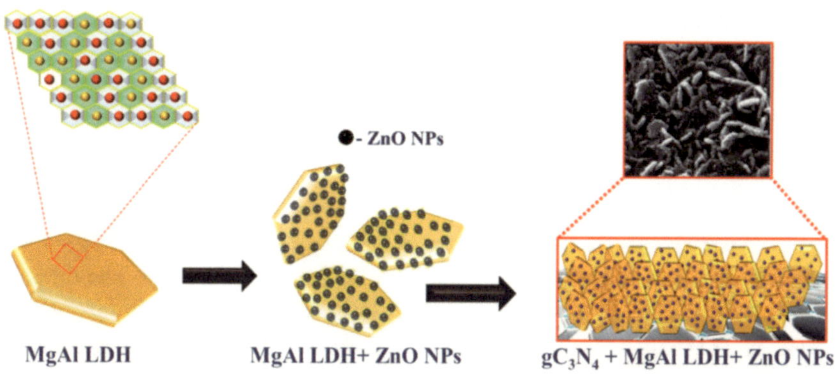

Figure 2.
Pictorial representation for the formation of g-C_3N_4\ZnO\Mg-Al LDH 2D/2D LDH tertiary nano-composite [34].

for 24 h. Later, the autoclave was cooled down naturally. Finally, the obtained solution was centrifuged and washed with DDW water and dried at 80°C for overnight to obtain the final product.

2.4 Preparation of 2D\2D hybrid

A certain amount of g-C_3N_4 was dispersed into 20 ml of DDW and ultrasonicated for 30 min. Subsequently, the ZnO\Mg-Al precursor solution was prepared and mixed with ultrasonicated g-C_3N_4 nanosheets. Again, the mixture was ultrasonicated for 30 min in a beaker to form a homogeneous suspension. After that, the reaction mixture was transferred into a 100 ml Teflon liner stainless-steel autoclave, followed by heating in an oven under 180°C for 24 h. After that, the autoclave was cooled down naturally to ambient temperature. Finally, the sample was centrifuged and washed with DDW water and dried at 80°C for overnight to obtain the final product. **Figure 2** shows the formation of g-C_3N_4\ZnO\Mg-Al LDH 2D\2D hybrid.

3. Characterization details

The photocatalytic activities mostly depend on the material nature, specific surface area, and light energy utilization ratio [35] etc., and for these reasons the crystallinity, functional group, surface area morphology and photophysical properties of the as-prepared samples were systematically investigated by various analytical techniques.

i. Structural investigation

ii. Morphology analysis

iii. Elemental analysis

iv. Photophysical investigation

v. Surface area investigation

3.1 Structural investigation

The crystalline phases have a significant influence on the photocatalytic activities [36]. So the phase purity and crystallite size of the synthesized samples were evaluated by X-ray diffractometer using Riguku MiniFlux-II diffractometer using Cu Kα radiation (λ = 1.540 46 Å). The crystalline nature of the prepared samples were investigated through XRD analysis and **Figure 3** displays the XRD pattern of Mg-Al LDH, gC_3N_4, ZnO and g-C_3N_4\ZnO\Mg-Al LDH ternary nanocomposite.

The XRD pattern of the Mg-Al LDH and g-C_3N_4, sample is in good agreement with the JCPDS card no: 35-0965 [37] and 87-1526 [38], respectively. The diffraction peaks indexed to (003), (006), (012), (015), (018), and (110) are the plane reflections of a typical hydrotalcite-like phase screening, sharp and symmetric basal (00 l) reflection of LDH. The XRD pattern of ternary nanocomposite consists of g-C_3N_4, ZnO and Mg-Al LDH diffraction peaks indicate the formation of the composite.

3.2 Morphological analysis

3.2.1 FE-SEM analysis

The surface morphology of the prepared samples was investigated by using FESEM and HRTEM analyses, respectively. The FE-SEM images were obtained by using Zeiss SUPRA-25 and the particle size and morphology of the prepared samples were analyzed by using HR-TEM – Jeol/JEM 2100, with LaB6 source.

Figure 4 shows the FESEM images of (a) Mg-Al LDH, (b) g-C_3N_4, (c) ZnO and (d) gC_3N_4\ZnO\Mg-Al LDH ternary nanocomposite samples. The Mg-Al LDH consists of plenty of two-dimensionally structured hexagonal LDHs matrix with a layer by layer assembly. The size of the hexagonal nanoflakes is approximately 200 nm, which indicate the successful exfoliation of a 2D layer.

3.2.2 HRTEM analysis

Furthermore, the hexagonal formed hydrotalcite-like particles were seen from the HRTEM analysis (**Figure 4d** and **e**) and it concurs well with the morphology

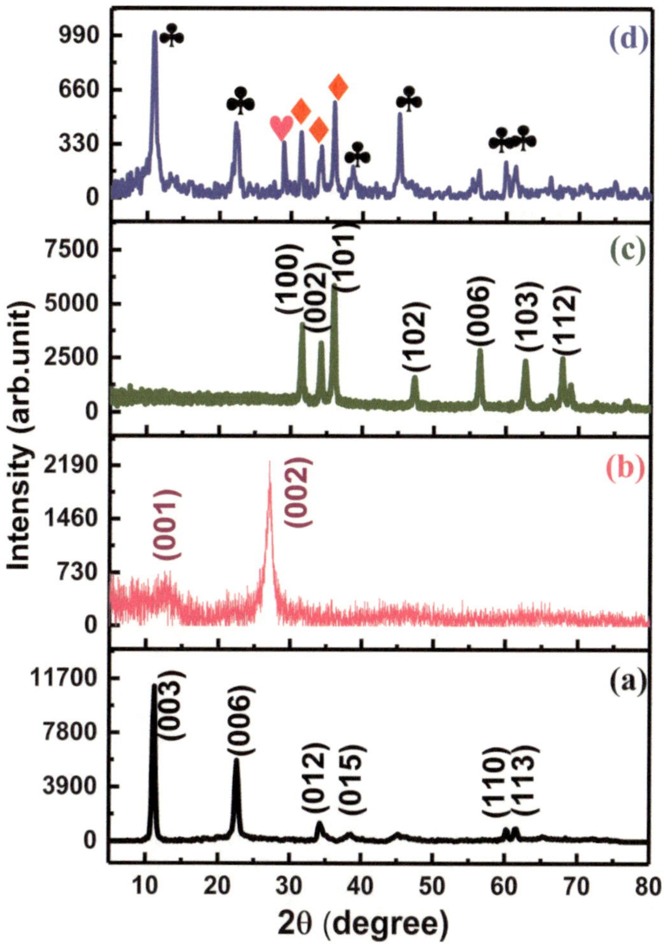

Figure 3.
XRD pattern of (a) Mg-Al LDH, (b) gC_3N_4, (c) ZnO and (d) g-C_3N_4\ZnO\Mg-Al LDH ternary nanocomposite [34].

acquired from FE-SEM investigation. Some dull spots showed up on the outside of the LDH, demonstrating that the ZnO nanoparticles are well attached on the LDH surfaces. Some dark spots appeared on the surface of the LDH, indicating that the ZnO nanoparticles are well attached to the surface of the as-prepared LDH. The 2D\2D ternary nanocomposites assembly was successfully obtained, and by arresting the ZnO\Mg-Al LDH sheets with g-C_3N_4 sheets, the formation of the 2D\2D ternary nanocomposite was possible. Surprisingly, after the formation of the ternary nanocomposite, the LDH loose its horizontal stacking arrangements and started aligning vertically on the surface of the g-C_3N_4 nanosheets. These types of arrangements provide a more active surface for the prepared photocatalysts.

3.3 Chemical analysis

The surface chemical composition of the prepared samples was confirmed by the FTIR, EDAX, Elemental mapping, and XPS analyses.

3.3.1 FTIR analysis

The vibrational bands of the prepared samples were analyzed by FTIR analysis (using a Bruker model Tensor 27 instrument) and the results are shown in **Figure 5a**. All the spectra, exhibit a strong band at 3700 to 3000 cm^{-1} which could be ascribed to the vibration of surface adsorbed water molecules and in the case of LDH plates, it is due to the formation of interlayer water molecules. Furthermore, several bands were observed in the 1200–1650 cm^{-1} region, which is assigned to the characteristic stretching modes of C—N heterocycles [39]. The absorption bands at 1620 cm^{-1} are associated with the C=O of the carboxylate groups. The occurrence of the feeble band at 1631 and 1643 cm^{-1} can be ascribed to the bending frequency and O—H asymmetric stretching vibration of the water molecules, respectively [40, 41]. The characteristic absorption band of ZnO samples was observed at 595 cm^{-1}, which is related to the metal-oxygen stretching vibration. The absorption

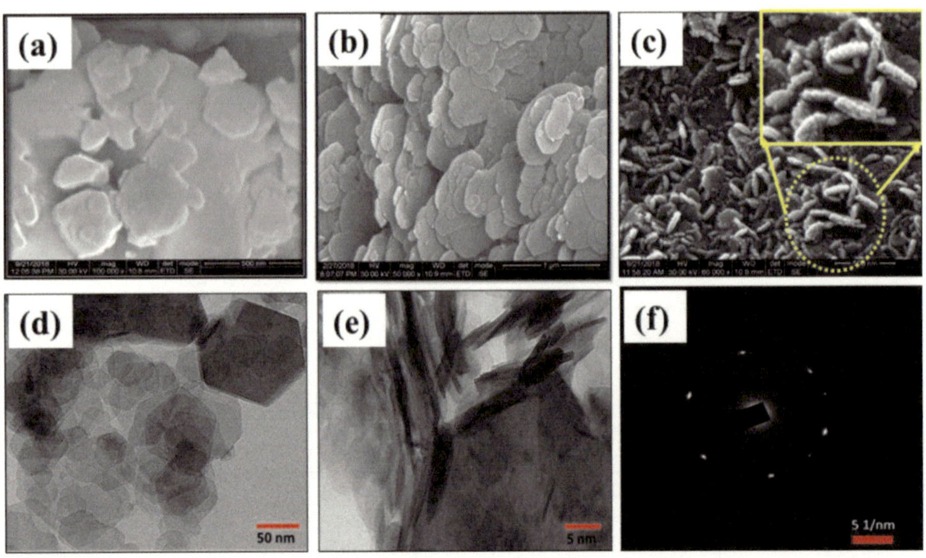

Figure 4.
Morphology analysis: FE-SEM images of (a) Mg-Al LDH, (b) g-C_3N_4, (c) g-C_3N_4\ZnO\Mg-Al LDH ternary nanocomposite and HRTEM images of (d) Mg-Al LDH, (e) g-C_3N_4\ZnO\Mg-Al LDH ternary nanocomposite, (f) SAED pattern of ternary nanocomposite [34].

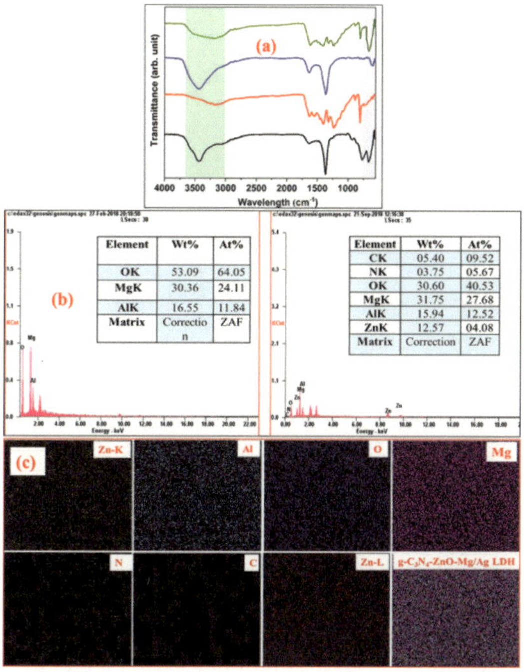

Figure 5.
(a) FTIR spectra for as prepared samples (b) EDX spectra of MgAl LDH and g-C_3N_4\ZnO\Mg-Al LDH ternary nanocomposite and (c) mapping analysis of g-C_3N_4\ZnO\Mg-Al LDH ternary nanocomposite [34].

bands at 653 cm^{-1} could be owed to the M—O—M lattice vibrations of the hexagonal sheets [42]. The successful intercalation of g-C_3N_4 and ZnO with Mg-Al LDH were observed from the presence of their corresponding bonds, in the FTIR results.

3.3.2 EDX analysis

Figure 5b shows the observed elemental composition of the ternary nanocomposite by EDX analysis (carried out using a JEOL Model JED 2300). From the EDX results, it could be able to observe the high percentages of O, Mg and Al elements, present in the as-prepared samples and no other impurities were observed. The elemental mapping of Mg-Al LDH and g-C_3N_4\ZnO\Mg-Al LDH ternary nanocomposite were presented in **Figure 5c**, which indicates the even distribution of the observed elements across the sample.

3.3.3 XPS analysis

XPS analysis was used to investigate the surface chemical composition of the prepared ternary nanocomposite and the obtained results were shown in **Figure 6**. The survey spectra show that the prepared sample is contain Mg, Al, Zn, O, C and N elements which creating peaks corresponding to Mg 1s, Al 2p, Zn 2p, O 1s, C 1s, and N 1s positions, respectively (**Figure 6A**). The high-resolution spectra of individual elements are presented in **Figure 6B**. The Mg^{2+} species are observed by the presence of Mg 2p, Mg 2s, Mg KLL, and Mg 1s state corresponding to the binding energies of 52.8, 90.8, 306.8, 351.8 and 1302.8 eV, respectively [43]. The high-resolution spectra of Mg 1s are fitted with three segments associating to the binding energies of 1307, 1308 and 1308.8 eV, which are ascribed to Mg, Mg-CO_3 and MgO [44, 45]. The Al attributed to the two states such as Al 2p and Al 1s and the characteristic

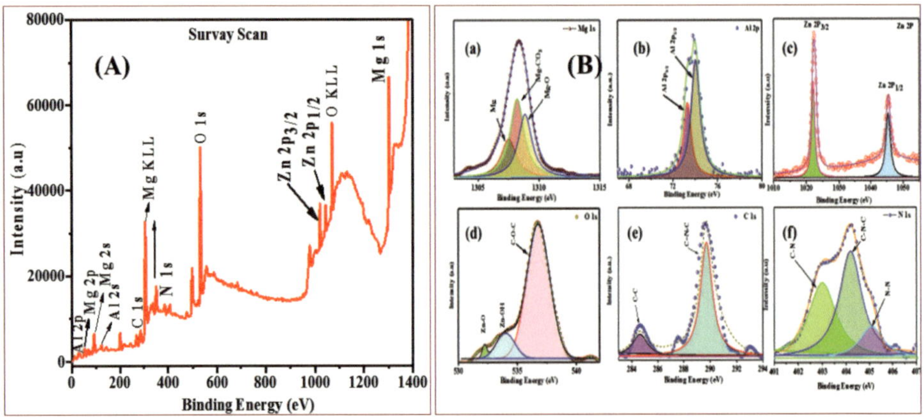

Figure 6.
XPS spectra of C_3N_4\ZnO\Mg-Al LDH ternary nanocomposite: (A) survey spectra and (B) high resolution XPS spectra of (a) Mg 1s (b) Al 2p (c) Zn 2p (d) O 1s (e) C 1s (f)N 1s [34].

peak were observed at 76.8 and 120.8 eV. The occurrence of Zn is seen from the two Zn 2p states as Zn $2P_{3\backslash 2}$ and Zn $2P_{1\backslash 2}$ corresponding to 1020.8 and 1043.8 eV respectively. It additionally uncovers that the Zn is available just in 2^+ oxidation state which affirms the conceivable bonding between Zn and O [10]. The oxygen O 1s is deconvoluted into three peaks corresponding to the O^{2-} at 532.2 eV, OH^- species at 533 eV and C—O—O at 536 eV, respectively [46]. In over-all, the inferior binding vitality of O 1s peaks emerges from the bond between O^{2-} and Zn^{2+} metal ions. The C 1s spectra can be deconvoluted into dual contributions such as 284.4 and 289 eV, assigned to the occurrence of sp^2 hybridized carbon atoms and C=*N*—C bonding [47], respectively. The N 1s spectra can be tailored into three basic peaks with the binding energies of 402, 404.1 and 405 eV, which are attributable to the binding of C—N, C—N—C, and N—N respectively [20]. Henceforth, the above observations affirm the formation of g-C_3N_4\ZnO\Mg-Al LDH ternary nanocomposite and the XPS results are in good agreement with FTIR, EDX and mapping analyses.

3.4 Photophysical investigation

3.4.1 UV-Vis absorption spectra

Optical properties possess a prominent role in the photocatalytic materials and therefore the photophysical properties of the prepared materials were investigated by UV-Vis and PL analyses. The optical absorption analysis was done using a SHIMADZU 3600 UV-Vis-NIR spectrophotometer and Emission spectrum of the as-prepared samples was recorded by using Horiba Jobin Yvon Spectro Fluromax 4. **Figure 7A** shows the UV-Vis. absorption spectra of the prepared samples. The absorption maxima were observed in the range between 320 and 450 nm. And the absorption of ternary nanocomposite was extended to the visible region and show an obvious red shift compared with the other samples, which may because of the interaction between the ZnO, LDH, and g-C_3N_4. The 2D\2D formation demonstrates a reality that the as-prepared ternary nanocomposite noticeable light vitality which can thusly create more charge transporters offered to contribution in the photocatalytic efficiency. Tauc's plot was used to determine the energy bandgap of the samples and the obtained values are 2.6, 3.5, 2.57 and 2.81 eV for LDH, ZnO,

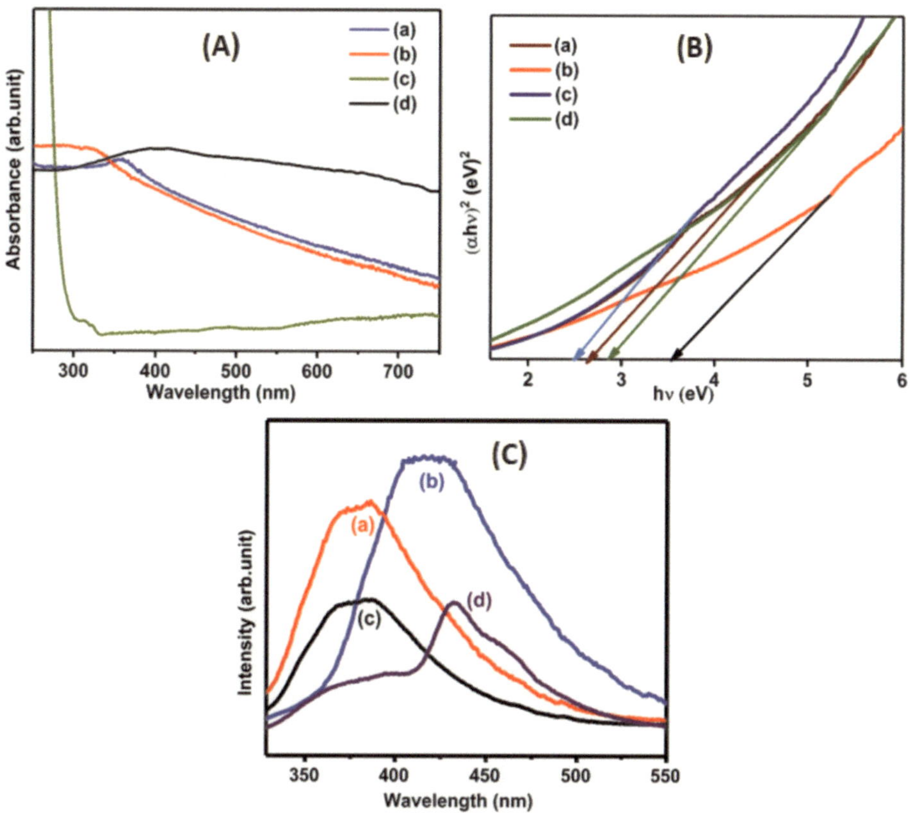

Figure 7.
(A) UV-Vis absorption spectra (B) Tauc's plots and (C) PL spectra of the prepared samples ((a) Mg-Al LDH and (b) ZnO (c) g-C_3N_4 (d) g-C_3N_4\ZnO\Mg-Al LDH ternary nanocomposite) [34].

g-C_3N_4, g-C_3N_4\ZnO\Mg-Al LDH ternary nanocomposite, respectively. The UV-Vis absorption results show the considerable enhancement in the visible light absorption and it is because of this reason, an enhancement in the photocatalytic performance of the as-prepared photocatalyst is observed (discussed in the latter part).

3.4.2 PL spectra

The emission spectrum is produced because of recombination of the charge carriers and it provides hints about the proficiency of charge carrier transformation, trapping, and separation of the photo generated electrons-holes pairs. The strong PL emission profile usually indicates the quick recombination of electron-hole pairs which provides low photocatalytic activity.

Figure 7C shows the PL emission spectra of (a) MgAl LDH, (b) g-C_3N_4, (c) ZnO and (d) g-C_3N_4\ZnO\Mg-Al LDH ternary nanocomposite, which were recorded using 320 nm as excitation wavelength. The LDH and ZnO nanoparticles exhibit a strong PL emission in the range from 350 to 450 nm. The pure g-C_3N_4 shows a strong emission about ~420 nm, which can be attributed to the fast electron-hole recombination process. It can be seen that, after the formation of ternary nanocomposite the emission was intensity was decreased which may due to the delocalization of electrons. In general, a decrease in the recombination rate gives rise to a low PL intensity, which results in the maximum photocatalytic activity.

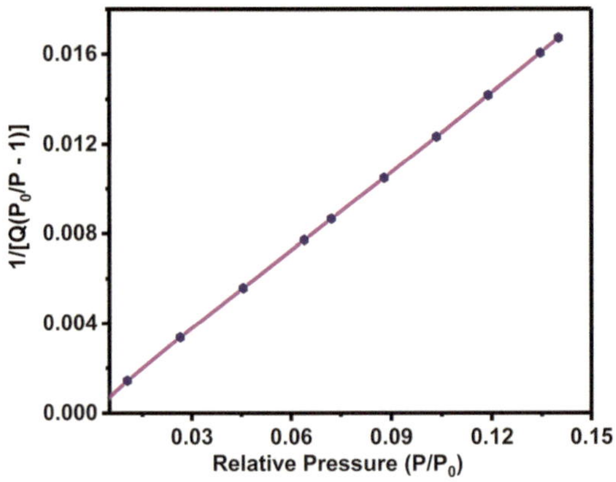

Figure 8.
BET surface area analysis of g-C_3N_4\ZnO\Mg-Al LDH ternary nanocomposite [34].

3.5 Surface area investigation

The specific surface area of the photocatalyst was determined by Brunauer-Emmett-Teller (BET) analysis through N_2 adsorption/desorption measurements at 25°C (**Figure 8**). The measured surface area of the ternary nanocomposite was ~37 $m^2\ g^{-1}$. The high surface area support more active species and reactants to be absorbed on its surface, which might proficiently help the kinetics of photo catalytic reaction.

4. Photocatalytic activity

4.1 Dye removal procedure

The photocatalytic activity of the as-prepared samples was investigated under UV-Vis light irradiation. The aqueous MB solution (20 mg L^{-1}) was prepared and kept in dark for 60 min to attain equilibrium. Later, 10 mg of the as-prepared photocatalyst was added to 10 ml MB solution and it was placed in a water jacketed photocatalytic reactor for the photocatalytic degradation process. A 250 W Hg lamp was used as the illumination source to excite the photocatalysts. In the whole reaction, the photocatalytic container was maintained at room temperature by circulating water. At 15 min time interval, 3 ml of solution was taken and centrifuged to remove the photocatalyst particles. The supernatant was examined by a Shimadzu UV3000 UV-Vis spectrophotometer and the dye absorption band maximum was observed at ~664 nm. The percentage of degradation was calculated using the Beer-Lambert relation [48]:

$$A(\lambda) = \log\left(I(\lambda)/I_0(\lambda)\right) = -\varepsilon l[BM] \quad (1)$$

where
A – absorbance at a given wavelength λ,
I_0 (λ) – incident light intensity,
I (λ) – light intensity transmitted through the MB solution,
ε – Molar attenuation coefficient of MB.
l – Path length of the beam of light.

The degradation efficiency was calculated by

$$\text{Degradation } (\%) = (C_0 - C)\backslash C_0 * 100 \qquad (2)$$

where,

C_0 is the initial dye concentration and
C is the dye concentration at time t from the start of the photocatalytic reaction.

For the reusability purpose, the as-prepared photocatalyst collected after the photocatalytic reaction by centrifuging, washed with DDW and then dried at 60°C.

4.2 Radical trapping experiment

To elucidate the reaction mechanism of the photocatalytic MB dye degradation, the radical trapping investigation was performed. In the scavenging activity, h^+, OH and O_2^- radicals are trapped by EDTA, 2-propanol and benzoquinone, respectively. The trapping experiments were carried out with the accumulation of different scavengers into the catalytic reaction. The reaction samples were taken from the photocatalytic reactor to record their UV-Vis absorption spectra.

4.3 Reaction mechanism of dye degradation

The photocatalytic activities of the Mg-Al LDH, g-C_3N_4, and g-C_3N_4\ZnO\Mg-Al LDH ternary nanocomposite were assessed under UV-Vis light illumination. In this work, MB dye was utilized as an objective contamination so as to decide the photocatalytic action of the impetuses under obvious light illumination. The MB dye solutions were prepared and the photocatalytic reactions were performed by adding the as-prepared samples to the MB dye solutions. The pure MB dye fragment shows a strong visible light absorption around 664 nm. The MB dye with prepared photocatalyst is subjecting under the visible light irradiation, corresponding absorption peak intensity was decreased, and the decreasing MB dye intensity is attributed to the degradation of MB dye through the photocatalytic activity. When increases the irradiation time, absorption intensity of MB dye molecules was decreased (i.e.) once increase the irradiation/reaction time, the large number of dye molecules can be degraded. In this process, a photocatalyst is irradiated by light with energy equal to or higher than the bandgap energy of the photocatalyst. This results in the excitation of an electron (e^-) from the valence band to the conduction band, leaving a hole (h^+) in the valence band. Before the recombination takes place, the photogenerated electrons (e^-) and holes (h^+) should be transferred to the surface of the photocatalyst in order to take part in the redox reactions with the adsorbed species. The redox reactions of electrons (e^-) and holes (h^+) with adsorbed oxygen and water molecules lead to the formation of superoxide radical anion ($^{\bullet}O_2^-$) and hydroxyl radical ($^{\bullet}OH$), respectively.

Among all the as-prepared photocatalyst samples, g-C_3N_4\ZnO\Mg-Al LDH ternary nanocomposite sample exhibit better photocatalytic activity. This could be attributed to a large number of electrons and holes generated by the as-prepared photocatalyst system, caused by the favorable visible light absorption. On the other hand, ZnO, a wide bandgap material, provides intermediate states to delay the electron–hole recombination, which could also contribute to the high photocatalytic activity. The morphological arrangements of the nanocomposite and its resultant electronic structure, (i.e.) the even distribution of ZnO intercalated LDH over the surface of g-C_3N_4 [49], collectively contribute to the effective separation of the photogenerated charge carriers. The observed photocatalytic degradation efficiencies of the as-prepared photocatalysts are 32%, 30%, 49% and 96.5% for ZnO, LDH, g-C_3N_4, and g-C_3N_4\ZnO\Mg-Al LDH ternary nanocomposite, respectively.

It has to be noted that the photocatalytic efficiency reported in the present study, betters our previous research work, in which ZnS QDs-LDH [49] exhibited a photocatalytic degradation efficiency of 95%. The enhanced photocatalytic efficiency is originated from the photocatalytic activity of N-rich g-C_3N_4, (i.e.,) the improved photocatalytic mechanism could be ascribed to the synergetic effect of graphitic N rich surface which offers more reactive sites for photocatalytic reaction. This in turn increases the utilization of the photo-separated charges towards the radical formation and corresponding degradation. Parallelly, the N-carbon acted as a co-catalyst to improve surface reaction kinetics and the nitrogen species directly contributed to the outstanding photocatalytic activity under visible light irradiation. Especially the nitrogen rich surface can achieve essential optical absorbance under visible light due to the mixing of O 2p states with p states. And also the nitrogen rich surface 2D/2D offers more active surface for the e^- transfer. The reaction kinetics of the MB dye degradation of the prepared photocatalysts is investigated by fitting the pseudo-first-order kinetic curve [50].

The plots of ln (C_0\C) against illumination time are appeared in **Figure 9**. From the kinetic graph, the ternary nanocomposite fits well and the outcome is in concurrence with the pseudo-first-order model. The impact of different scavengers on the photodegradation of MB dye solution was studied in order to recognize the role of receptive oxidative species in the photodegradation process. The role of H^+, OH and O_2 radicals were done individually, utilizing EDTA, 2-propanol, and BQ respectively. During the addition of EDTA and BQ, there were conspicuous variations in the photocatalytic process, which shows that H^+ and O_2 radicals are influences in MB dye degradation. But after the addition of 2-Proponal (scavenger for the OH radical), the degradation of MB is highly suppressed than other reactions, indicating the major of *OH in MB dye degradation. From this result, it is clear that the photocatalytic degradation process is led by the contribution of hydroxyl radical (*OH). After the addition of the as-prepared photocatalyst into the reaction and irradiating with visible light, the electrons were photoexcited from the valence band (VB) to the conduction band (CB). Once electrons are excited, the hole act as an oxidizing agent and oxidize the aquatic or the dye directly to form *OH radicals.

These OH reactive species are responsible for the efficient degradation of organic pollutants in water. The following equation represent the possible photocatalytic reaction mechanism of MB dye degradation under visible light irradiation.

$$
\begin{aligned}
&g - C_3N_4/ZnO/MgAl(LDH) + h\nu \rightarrow g - C_3N_4/MgAlLDH\left(h^+ + e^-\right)ZnO \\
&g - C_3N_4/MgAl(LDH) + \left(h^+\right) \rightarrow (e^-)(free - electrons) \\
&e^- + O_2 \rightarrow O_2^{\bullet} \\
&h^+ + OH^- \rightarrow OH \\
&h^+ + H_2O \rightarrow OH + H^+ \\
&O_2^{\bullet} + MB \rightarrow CO_2 + H_2O(By - product) \\
&OH^{\bullet} + MB \rightarrow CO_2 + H_2O(By - product) \\
&h^+ + MB \rightarrow CO_2 + H_2O(By - product)
\end{aligned}
\tag{3}
$$

The **Figure 10** shows the schematic illustration of possible photocatalytic degradation of MB dye under visible light irradiation. The reusability of the prepared photocatalyst was studied by performing continual tests under same reaction conditions (shown in **Figure 11b**). The fresh MB solution was utilized for resulting cycles. Subsequently in each cycle, the prepared catalyst was isolated from the photocatalytic reactor through centrifugation. After 4 cycles, the degradation ability of the prepared catalyst was slightly reduced and it might be because of the loss

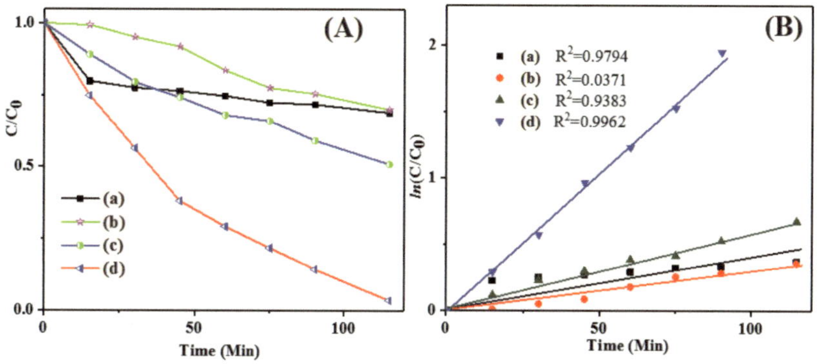

Figure 9.
(A) Photocatalytic degradation and (B) pseudo-first-order kinetics for the degradation of MB over (a) ZnO (b) LDH(c) g-C_3N_4 (d) g-C_3N_4\ZnO\Mg-Al LDH ternary nanocomposite [34].

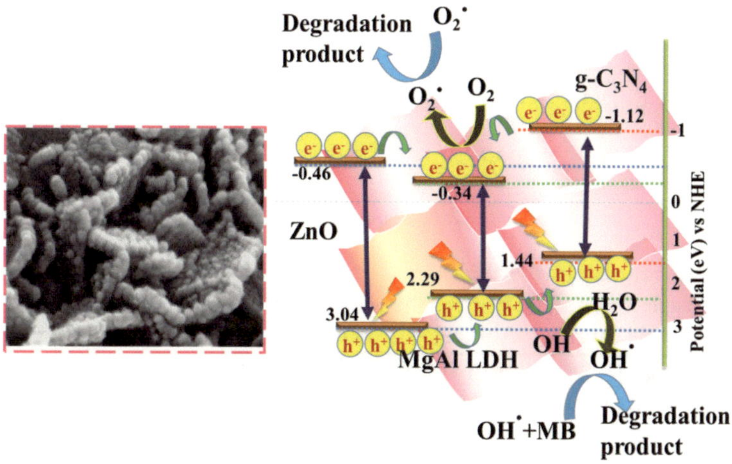

Figure 10.
Schematic illustration of proposed photocatalytic reaction mechanism.

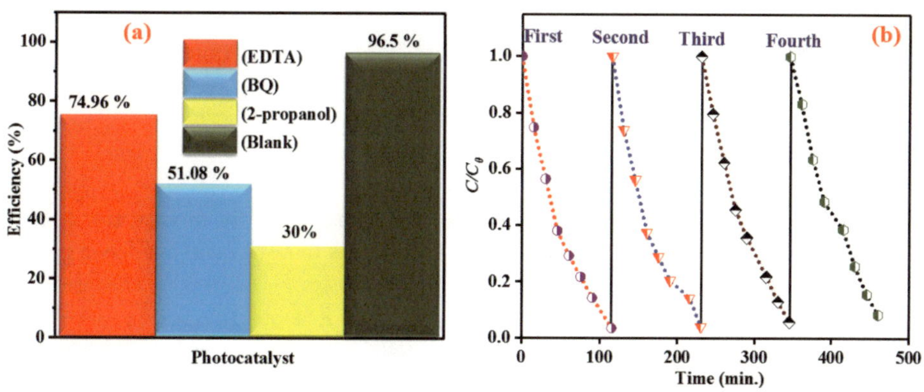

Figure 11.
(a) Radical trapping experiments of active species over g-C_3N_4\ZnO\Mg-Al LDH ternary nanocomposite and (b) reusability of g-C_3N_4\ZnO\Mg-Al LDH ternary nanocomposite in the photodegradation of MB [34].

of catalyst during the recycling process. The photocatalytic dye degradation activity of the prepared sample was compared to the previously reported nanomaterials, which is given in **Table 1**.

S. No	Catalyst	Dosage of dye (mg)	Dye	Light source	Catalyst efficiency (%)	Reaction time (min)	References
1.	PANI-ZnO	100	MB	250 W	98.3	180	[51]
2.	Nitrogen doped dual phase titanate	10	MB	450 W	~97	310	[52]
3.	Zirconia	50	MB	500 W	~78	300	[53]
4.	CA-CNT/TiO_2-NH_2	2 cm × 13 cm (CNT)	MB	40 W	80	300	[54]
5.	TiO_2	100	MB	6 W	97	180	[55]
6.	g-C_3N_4\ZnO\Mg-Al LDH	10	MB	250 W	96.5	120	Present work

Table 1.
Comparison table of MB dye degradation using different photocatalyst with degradation (%) of previously reported nanomaterials.

Form the experimental results, it was confirmed that the as-prepared ternary nanocomposite exhibits remarkable photocatalytic activity and reusability under visible light to photo-degrade MB dye.

5. Conclusion

In summary, the hydrothermally prepared 2D\2D ternary nanocomposite was used as an efficient photocatalyst for the photodegradation of MB dye. The nitrogen-rich 2D\2D (g-C_3N_4 and Mg-Al LDH) surface significantly enhanced the photocatalytic efficiency under the visible light irradiation due to the improved photo active surfaces. Especially, in ternary nanocomposite Mg-Al LDH 2D nanoplates are vertically well aligned on the surface of the g-C_3N_4 2D nanosheets. This 2D/2D arrangement results effectively enhances the photocatlytic activity due to the efficient separation of photo-induced charge carriers and transfer by the incorporation of ZnO into LDH brucite layers. In addition, the g-C_3N_4 surface contributed to the efficient charge injection in the photocatalytic reaction. The novel g-C_3N_4\ZnO\Mg-Al LDH ternary nanocomposite can be used as a proficient material for the photocatalytic degradation of MB dyes under visible light irradiation.

6. Future aspects/prospective

1. The as prepared ternary nanocomposite show large specific surface area and efficiently active in the visible light region so this material can used as a good photocatalytic for other organic dye degradation

2. The LDH have large number of interlayer galleries which can capable to adsorb the pollutants such as heavy metals synthetic or organic dyes.

Author details

Kandasamy Bhuvaneswari[1], Thangavelu Pazhanivel[1*], Govindasamy Palanisamy[1] and Ganapathi Bharathi[2]

1 Smart Materials Interface Laboratory, Department of Physics, Periyar University, Salem, Tamilnadu, India

2 Key Laboratory of Optoelectronic Devices and Systems of Guangdong Province, College of Optoelectronic Engineering, Shenzhen University, Shenzhen, Guangdong Province, P.R. China

*Address all correspondence to: pazhanit@gmail.com

References

[1] Axelson SO. IARC monographs on the evaluation of carcinogenic risks to humans. IARC Monographs on the Evaluation of Carcinogenic Risks to Humans. 2010;**94**:9-38

[2] Sun W, Meng S, Zhang S, Zheng X, Ye X, Fu X, et al. Insight into the transfer mechanisms of Photogenerated carriers for heterojunction photocatalysts with the analogous positions of valence band and conduction band: A case study of ZnO/TiO_2. Journal of Physical Chemistry C. 2018;**122**:15409-15420. DOI: 10.1021/acs.jpcc.8b03753

[3] Prabhu S, Pudukudy M, Sohila S, Harish S, Navaneethan M, Navaneethan D. Synthesis, structural and optical properties of ZnO spindle/reduced graphene oxide composites with enhanced photocatalytic activity under visible light irradiation. Optical Materials. 2018;**79**:186-195. DOI: 10.1016/j.optmat.2018.02.061

[4] Nakata K, Fujishima A. TiO_2 photocatalysis: Design and applications. Journal of Photochemistry and Photobiology C Photochemistry Reviews. 2012;**13**:169-189. DOI: 10.1016/j.jphotochemrev.2012.06.001

[5] Prakash K, Senthil Kumar P, Pandiaraj S, Saravanakumar K, Karuthapandian S. Controllable synthesis of SnO_2 photocatalyst with superior photocatalytic activity for the degradation of methylene blue dye solution. Journal of Experimental Nanoscience. 2016;**11**:1138-1155. DOI: 10.1080/17458080.2016.1188222

[6] Kim SP, Choi MY, Choi HC. Photocatalytic activity of SnO_2 nanoparticles in methylene blue degradation. Materials Research Bulletin. 2016;**74**:85-89. DOI: 10.1016/j.materresbull.2015.10.024

[7] Lee KM, Lai CW, Ngai KS, Juan JC. Recent developments of zinc oxide based photocatalyst in water treatment technology: A review. Water Research. 2016;**88**:428-448. DOI: 10.1016/j.watres.2015.09.045

[8] Guo M, Diao P, Cai S. Hydrothermal growth of perpendicularly oriented ZnO nanorod array film and its photoelectrochemical properties. Applied Surface Science. 2005;**249**:71-75. DOI: 10.1016/j.apsusc.2004.11.053

[9] Jalal R, Goharshadi EK, Abareshi M, Moosavi M, Yousefi A, Nancarrow P. ZnO nanofluids: Green synthesis, characterization, and antibacterial activity. Materials Chemistry and Physics. 2010;**121**:198-201. DOI: 10.1016/j.matchemphys.2010.01.020

[10] Thirumalai K, Shanthi M, Swaminathan M. Hydrothermal fabrication of natural sun light active Dy_2WO_6 doped ZnO and its enhanced photo-electrocatalytic activity and self-cleaning properties. RSC Advances. 2017;**7**:7509-7518. DOI: 10.1039/c6ra24843h

[11] Raza W, Faisal SM, Owais M, Bahnemann D, Muneer M. Facile fabrication of highly efficient modified ZnO photocatalyst with enhanced photocatalytic, antibacterial and anticancer activity. RSC Advances. 2016;**6**:78335-78350. DOI: 10.1039/c6ra06774c

[12] Yuan G, Xiang J, Jin H, Wu L, Jin Y, Zhao Y. Anchoring ZnO nanoparticles in nitrogen-doped graphene sheets as a high-performance anode material for lithium-ion batteries. Materials. 2018;**11**:15-17. DOI: 10.3390/ma11010096

[13] Zhang L, Li L, Sun X, Liu P, Yang D, Zhao X. ZnO-layered double hydroxide@graphitic carbon nitride composite for consecutive adsorption

and photodegradation of dyes under UV and visible lights. Materials. 2016;**9**:11. DOI: 10.3390/ma9110927

[14] Zhao Y, Zhao F, Wang X, Xu C, Zhang Z, Shi G, et al. Graphitic carbon nitride nanoribbons: Graphene-assisted formation and synergic function for highly efficient hydrogen evolution. Angewandte Chemie, International Edition. 2014;**53**:13934-13939. DOI: 10.1002/anie.201409080

[15] Xu C, Han Q, Zhao Y, Wang L, Li Y, Qu L. Sulfur-doped graphitic carbon nitride decorated with graphene quantum dots for an efficient metal-free electrocatalyst. Journal of Materials Chemistry A. 2015;**3**:1841-1846. DOI: 10.1039/c4ta06149g

[16] Xu M, Chai B, Yan J, Wang H, Ren Z, Paik K-W. Facile synthesis of fluorine doped graphitic carbon nitride with enhanced visible light photocatalytic activity. Nano. 2016;**11**:1650137. DOI: 10.1142/S179329201650137X

[17] Yuan B, Chu Z, Li G, Jiang Z, Hu T, Wang Q, et al. Water-soluble ribbon-like graphitic carbon nitride (g-C_3N_4): Green synthesis, self-assembly and unique optical properties. Journal of Materials Chemistry C. 2014;**2**: 8212-8215. DOI: 10.1039/c4tc01421a

[18] Patnaik S, Sahoo DP, Parida K. An overview on Ag modified g-C_3N_4 based nanostructured materials for energy and environmental applications. Renewable and Sustainable Energy Reviews. 2018; **82**:1297-1312. DOI: 10.1016/j.rser.2017.09.026

[19] Majeed I, Manzoor U, Kanodarwala FK, Nadeem MA, Nadeem MA, Hussain E, et al. Pd-Ag decorated g-C_3N_4 as an efficient photocatalyst for hydrogen production from water under direct solar light irradiation. Catalysis Science & Technology. 2018;**8**:1183-1193. DOI: 10.1039/c7cy02219k

[20] Han Q, Hu C, Zhao F, Zhang Z, Chen N, Qu L. One-step preparation of iodine-doped graphitic carbon nitride nanosheets as efficient photocatalysts for visible light water splitting. Journal of Materials Chemistry A. 2015;**3**: 4612-4619. DOI: 10.1039/c4ta06093h

[21] Zheng Y, Liu J, Liang J, Jaroniec M, Qiao SZ. Graphitic carbon nitride materials: Controllable synthesis and applications in fuel cells and photocatalysis. Energy & Environmental Science. 2012;**5**: 6717-6731. DOI: 10.1039/c2ee03479d

[22] Deonikar VG, Koteshwara Reddy K, Chung WJ, Kim H. Facile synthesis of Ag_3PO_4/g-C_3N_4 composites in various solvent systems with tuned morphologies and their efficient photocatalytic activity for multi-dye degradation. Journal of Photochemistry and Photobiology A: Chemistry. 2019;**368**:168-181. DOI: 10.1016/j.jphotochem.2018.09.034

[23] Liu T, Wang L, Lu X, Fan J, Cai X, Gao B, et al. Comparative study of the photocatalytic performance for the degradation of different dyes by $ZnIn_2S_4$: Adsorption, active species, and pathways. RSC Advances. 2017;**7**: 12292-12300. DOI: 10.1039/c7ra00199a

[24] Liang Q, Li Z, Huang ZH, Kang F, Yang QH. Holey graphitic carbon nitride nanosheets with carbon vacancies for highly improved photocatalytic hydrogen production. Advanced Functional Materials. 2015;**25**: 6885-6892. DOI: 10.1002/adfm.201503221

[25] Shi L, Li DQ, Li SF, Wang JR, Evans DG, Duan X. Structure, flame retarding and smoke suppressing properties of Zn-Mg-Al-CO_3 layered double hydroxides. Chinese Science Bulletin. 2005;**50**:1101-1104. DOI: 10.1360/04wb0043

[26] Jabeen M, Ishaq M, Song W, Xu L, Deng Q. Synthesis of Ni/Co/Al-layered

triple hydroxide@brominated graphene hybrid on nickel foam as electrode material for high-performance supercapacitors. RSC Advances. 2017;7: 46553-46565. DOI: 10.1039/c7ra08744f

[27] Zhang F, Du N, Li H, Liang X, Hou W. Sorption of Cr(vi) on Mg-Al-Fe layered double hydroxides synthesized by a mechanochemical method. RSC Advances. 2014;**4**:46823-46830. DOI: 10.1039/c4ra07553f

[28] Hobbs C, Jaskaniec S, McCarthy EK, Downing C, Opelt K, Güth K, et al. Structural transformation of layered double hydroxides: An in situ TEM analysis. npj 2D Materials and Applications. 2018;**2**:1-10. DOI: 10.1038/s41699-018-0054-6

[29] Mohapatra L, Parida K. A review on the recent progress, challenges and perspective of layered double hydroxides as promising photocatalysts. Journal of Materials Chemistry A. 2016; **4**:10744-10766. DOI: 10.1039/c6ta01668e

[30] Qu J, He X, Li X, Ai Z, Li Y, Zhang Q, et al. Precursor preparation of Zn-Al layered double hydroxide by ball milling for enhancing adsorption and photocatalytic decoloration of methyl orange. RSC Advances. 2017;7: 31466-31474. DOI: 10.1039/c7ra05316a

[31] Dvininov E, Ignat M, Barvinschi P, Smithers MA, Popovici E. New SnO2/MgAl-layered double hydroxide composites as photocatalysts for cationic dyes bleaching. Journal of Hazardous Materials. 2010;**177**:150-158. DOI: 10.1016/j.jhazmat.2009.12.011

[32] Seftel EM, Mertens M, Cool P. The influence of the Ti4+ location on the formation of self-assembled nanocomposite systems based on TiO_2 and Mg/Al-LDHs with photocatalytic properties. Applied Catalysis B: Environmental. 2013;**134–135**:274-285. DOI: 10.1016/j.apcatb.2013.01.032

[33] Dutta K, Das S, Pramanik A. Concomitant synthesis of highly crystalline Zn-Al layered double hydroxide and ZnO: Phase interconversion and enhanced photocatalytic activity. Journal of Colloid and Interface Science. 2012;**366**: 28-36. DOI: 10.1016/j.jcis.2011.09.081

[34] Bhuvaneswari K, Palanisamy G, Pazhanivel T, Maiyalagan T, Bharathi G. Photodegradation activity of nitrogen-rich graphitic carbon nitride intercalated ZnO\Mg-Al layered double hydroxide ternary nanocomposites on methylene blue dye. ChemistrySelect. 2019;**4**: 2982-2990. DOI: 10.1002/slct.201900146

[35] Saravanan R, Aviles J, Gracia F, Mosquera E, Kumar V. International journal of biological macromolecules crystallinity and lowering band gap induced visible light photocatalytic activity of TiO_2/CS (chitosan) nanocomposites. International Journal of Biological Macromolecules. 2018;**109**: 1239-1245. DOI: 10.1016/j.ijbiomac.2017.11.125

[36] Li W, Wang X, Wang Z, Meng Y, Sun X, Yan T, et al. Relationship between crystalline phases and photocatalytic activities. Materials Research Bulletin. 2016;**83**:259-267. DOI: 10.1016/j.materresbull.2016.06.021

[37] Liu J, Lv G, Gu W, Li Z, Tang A, Mei L. A novel luminescence probe based on layered double hydroxides loaded with quantum dots for simultaneous detection of heavy metal ions in water. Journal of Materials Chemistry C. 2017;5:5024-5030. DOI: 10.1039/c7tc00935f

[38] Qi Y, Liang Q, Lv R, Shen W, Kang F, Huang Z. Synthesis and photocatalytic activity of mesoporous g-C_3N_4/MoS_2 hybrid catalysts. Royal Society Open Science. 2018:2-10

[39] Tonda S, Kumar S, Bhardwaj M, Yadav P, Ogale S. G-C_3N_4/NiAl-LDH

2D/2D hybrid heterojunction for high-performance photocatalytic reduction of CO_2 into renewable fuels. ACS Applied Materials & Interfaces. 2017;**10**: 2667-2678. DOI: 10.1021/acsami.7b18835

[40] Chen J, Sheng Y, Song Y, Chang M, Zhang X, Cui L, et al. Multimorphology mesoporous silica nanoparticles for dye adsorption and multicolor luminescence applications. ACS Sustainable Chemistry & Engineering. 2018;**6**: 3533-3545. DOI: 10.1021/acssuschemeng.7b03849

[41] Sobhana L, Sarakha M, Prevot V, Fardim P. Layered double hydroxides decorated with Au-Pd nanoparticles to photodegradate Orange II from water. Applied Clay Science. 2016;**134**:120-127. DOI: 10.1016/j.clay.2016.06.019

[42] Indrasekara S, Kottegoda N. Synthesis and characterisation of exfoliated layered double hydroxide (LDH)/sugar nanocomposites. Journal of the National Science Foundation of Sri Lanka. 2011;**39**:113-119. DOI: 10.4038/jnsfsr.v39i2.3171

[43] Wysocka J, Krakowiak S, Ryl J. Evaluation of citric acid corrosion inhibition efficiency and passivation kinetics for aluminium alloys in alkaline media by means of dynamic impedance monitoring. Electrochimica Acta. 2017; **258**:1463-1475. DOI: 10.1016/j.electacta.2017.12.017

[44] Tan XF, Liu YG, Gu YL, Liu SB, Zeng GM, Cai X, et al. Biochar pyrolyzed from MgAl-layered double hydroxides pre-coated ramie biomass (Boehmeria nivea (L.) gaud.): Characterization and application for crystal violet removal. Journal of Environmental Management. 2016;**184**: 85-93. DOI: 10.1016/j.jenvman.2016.08.070

[45] Vinayan BP, Zhao-Karger Z, Diemant T, Chakravadhanula VSK, Schwarzburger NI, Cambaz MA, et al. Performance study of magnesium-sulfur battery using a graphene based sulfur composite cathode electrode and a non-nucleophilic Mg electrolyte. Nanoscale. 2016;**8**:3296-3306. DOI: 10.1039/c5nr04383b

[46] Xu J, Gai S, He F, Niu N, Gao P, Chen Y, et al. A sandwich-type three-dimensional layered double hydroxide nanosheet array/graphene composite: Fabrication and high supercapacitor performance. Journal of Materials Chemistry A. 2014;**2**:1022-1031. DOI: 10.1039/c3ta14048b

[47] Shinde SS, Sami A, Lee JH. Electrocatalytic hydrogen evolution using graphitic carbon nitride coupled with nanoporous graphene co-doped by S and Se. Journal of Materials Chemistry A. 2015;**3**:12810-12819. DOI: 10.1039/c5ta02656c

[48] Petit M, Michez L, Raimundo JM, Malinowski T, Dumas P. An introduction to photocatalysis through methylene blue photodegradation. European Journal of Physics. 2016;**37**: 1-11. DOI: 10.1088/0143-0807/37/6/065808

[49] Bhuvaneswari K, Palanisamy G, Pazhanivel T, Bharathi G, Nataraj D. Photocatalytic performance on visible light induced ZnS QDs-MgAl layered double hydroxides hybrids for methylene blue dye degradation. ChemistrySelect. 2018;**3**:13419-13426. DOI: 10.1002/slct.201803183

[50] Moussa H, Chouchene B, Mozet K, Schneider R, Gries T, Medjahdi G, et al. Growth of ZnO nanorods on graphitic carbon nitride gCN sheets for the preparation of photocatalysts with high visible-light activity. ChemCatChem. 2018;**10**:1-12. DOI: 10.1002/cctc.201801206

[51] Saravanan R, Sacari E, Gracia F, Khan MM, Mosquera E, Gupta VK.

Conducting PANI stimulated ZnO system for visible light photocatalytic degradation of coloured dyes. Journal of Molecular Liquids. 2016;**221**:1029-1033. DOI: 10.1016/j.molliq.2016.06.074

[52] Cheng YH, Huang Y, Kanhere PD, Subramaniam VP, Gong D, Zhang S, et al. Dual-phase titanate/Anatase with nitrogen doping for enhanced degradation of organic dye under visible light. Chemistry - A European Journal. 2011;**17**:2575-2578. DOI: 10.1002/chem.201002256

[53] Teeparthi SR, Awin EW, Kumar R. Dominating role of crystal structure over defect chemistry in black and white zirconia on visible light photocatalytic activity. Scientific Reports. 2018;**8**:1-11. DOI: 10.1038/s41598-018-23648-0

[54] Salama A, Mohamed A, Aboamera NM, Osman TA, Khattab A. Photocatalytic degradation of organic dyes using composite nanofibers under UV irradiation. Applied Nanoscience. 2018;**8**:155-161. DOI: 10.1007/s13204-018-0660-9

[55] Azeez F, Al-Hetlani E, Arafa M, Abdelmonem Y, Nazeer AA, Amin MO, et al. The effect of surface charge on photocatalytic degradation of methylene blue dye using chargeable titania nanoparticles. Scientific Reports. 2018;**8**:1-9. DOI: 10.1038/s41598-018-25673-5